MW01517717

WORKING IN
HIGH RISK ENVIRONMENTS

GEORGIAN COLLEGE LIBRARY

GEOR-BK
48.34

WORKING IN HIGH RISK ENVIRONMENTS

Developing Sustained Resilience

Edited by

DOUGLAS PATON, Ph.D.

School of Psychology, University of Tasmania
Launceston, Tasmania, Australia

and

JOHN M. VIOLANTI, Ph.D.

School of Public Health and Health Professions
Department of Social & Preventive Medicine
State University of New York at Buffalo, NY

CHARLES C THOMAS • PUBLISHER, LTD.
Springfield • Illinois • U.S.A.

Library Commons
Georgian College
825 Memorial Avenue
Box 2316
Orillia, ON L3V 6S2

Published and Distributed Throughout the World by

CHARLES C THOMAS • PUBLISHER, LTD.
2600 South First Street
Springfield, Illinois 62704

This book is protected by copyright. No part of
it may be reproduced in any manner without written
permission from the publisher. All rights reserved.

© 2011 by CHARLES C THOMAS • PUBLISHER, LTD.

ISBN 978-0-398-08692-3 (hard)
ISBN 978-0-398-08693-0 (paper)
ISBN 978-0-398-08694-7 (ebook)

Library of Congress Catalog Card Number: 2011025783

With THOMAS BOOKS *careful attention is given to all details of manufacturing
and design. It is the Publisher's desire to present books that are satisfactory as to their
physical qualities and artistic possibilities and appropriate for their particular use.*
THOMAS BOOKS *will be true to those laws of quality that assure a good name
and good will.*

Printed in the United States of America
MM-R-3

Library of Congress Cataloging-in-Publication Data

Working in high risk environments : developing sustained resilience
/ edited by Douglas Paton and John M. Violanti.
  p. cm.
Includes bibliographical references and index.
ISBN 978-0-398-08692-3 (hard) -- ISBN 978-0-398-08693-0 (pbk.) --
ISBN 978-0-398-08694-7 (ebook)
1. Job stress. 2. Hazardous occupations--Psychological aspects. 3.
Emergency management--Psychological aspects. 4. Resilience
(Personality trait). 5. Stress management. I. Paton, Douglas. II.
Violanti, John M. III. Title.

HF5548.85.W678 2011
363.1101'9--dc23
                                                      2011025783

CONTRIBUTORS

Jeff Ayton commenced with the Australian Antarctic Division as Chief Medical Officer in 2002 with responsibility for the Australian Antarctic program medical support and leadership of human biology and medicine research. He is Australian Delegate to the Scientific Committee of Antarctic Research Life Sciences Standing Scientific Group and deputy chief officer of the SCAR LSSSG Expert Group of Human Biology and Medicine. In 1992, Jeff wintered and conducted research at Casey Station, Antarctica, as a remote area general practitioner. His research interests include Antarctic medicine and the human physiological and psychology responses to Antarctic sojourns and their translation to other remote and extreme environments. Jeff is the current President of the Australian College of Rural and Remote Medicine.

Paul T. Bartone is a Senior Research Fellow at the Center for Technology and National Security Policy, National Defense University. Prior to retiring from the U.S. Army, Colonel Bartone was the senior Army research psychologist and Consultant to the Surgeon General and the Assistant Corps Chief for Medical Allied Sciences. Bartone is a Fellow of the American Psychological Association (APA), past President of APA's Division 19, the Society for Military Psychology, and a charter member of the Association for Psychological Science. He has written extensively on various topics related to leadership, stress, health, and adaptation.

Dr. Charles Barry is a Senior Research Fellow with the Center for Technology and National Security Policy (CTNSP). He holds undergraduate and graduate degrees in international relations and earned his Doctor of Public Administration at the University of Baltimore, concentrating in the field of Information Resource Management. He is a member of the Pi Alpha Alpha National Honor Society in Public Administration, the Association of the U.S. Army, the Army Aviation Association of America, and the Military Officers Association. He has been a Woodrow Wilson Foundation Fellow since 2004. Dr. Barry is a retired career soldier with combat leadership service in infantry and aviation. In 12 years overseas, he experienced numerous deployments in Asia and Europe as well as the Caribbean, Central America, and Africa. He also served nearly 10 years as a joint strategic planner in Europe and Washington.

Cherie Castellano is the director of several nationally renowned peer support programs at University Behavioral HealthCare–University of Medicine & Dentistry of New Jersey. Cop 2 Cop, Vet 2 Vet, and Mom 2 Mom are peer-based helpline programs utilizing "reciprocal peer support" developed by Ms. Castellano for a model of success. Cherie is married to a law enforcement professional and has two sons whom she considers her greatest achievements.

Emma E. Doyle is a postdoctoral fellow funded by N.Z's Foundation for Research, Science, and Technology and based at the Joint Centre for Disaster Research at Massey University, Wellington, New Zealand, working on the project "Effective Integration of Science into Emergency Decision-Making Processes." Emma's interests lie at the interface between physical science and emergency management, with a primary focus on the critical decisions made during a natural hazard event, including evacuations and hazard zone limitations. Her current research involves an assessment of the scientific needs of decision makers during a natural hazard event and investigations into how these critical decisions are influenced by the presentation of the scientific information, forecasts, and model outputs. This has involved the development of an Emergency Communications Research Laboratory and the analysis of two NZ national disaster exercises, Capital Quake (earthquake) and Exercise Ruaumoko (volcano). Previous research includes a postdoctoral position at Massey University, Palmerston North, working with the Volcanic Risk Solutions group on the Marsden-funded project "Capturing the Secrets of a Life-Size Lahar"; a PhD in volcanology at Bristol University, UK, in 2008, specialising in modelling hazardous volcanic flows; and a Master's by Research investigating volcanic eruption precursors at Leeds University, UK, in 2003.

Dr. James J. Drylie is the Executive Director of the School of Criminal Justice & Public Administration at Kean University of New Jersey. He has conducted exhaustive research on the subject of suicide-by-cop formulating the theory of victim-scripted suicide as an objective measure in examining these types of police-involved shootings and is regualrly consulted on matters of this nature. Retired at the rank of captain, after 25 years of service with a large suburban New Jersey police department, he routinely lectures on the subject of suicide-by-cop and related issues to a variety of academic and practitioner audiences.

During the course of his law enforcement career, Dr. Drylie was recognized as an expert in various aspects related to the use of force, specifically deadly force, and was responsible for training countless recruits and police officers throughout the New York/New Jersey Region. Dr. Drylie has recently begun working with clinicians who study the impact of large-scale disaster on first responders and is researching the links between the psychological truama from terrorist incidents, natural and man-made disaster, and human-caused incidents such as police shootings. He presented his work, *Cultural Boundaries and Suicide Terrorism,* at the 2010 International Symposium on Terrorism and Transnational Crime in Antalya, Turkey. Dr. Drylie received his PhD in Criminal Justice from the City University of New York, John Jay College of Criminal Justice. He has published his dissertation work in *Suicide by Cop: Scripted Behavior Resulting in Police Deadly Force* (2007) as well as coauthoring a textbook, *Cop-I-Cide: Concepts, Cases and Controversies of Suicide by Cop* (2008) with the noted expert on police suicide, Dr. John Violanti.

George S. Everly, Jr. is Executive Director of Resiliency Science Institutes at University of Maryland, Baltimore County (UMBC) Training Centers; Professor of Psychology, Loyola University Maryland; Associate Professor of Psychiatry at the Johns Hopkins University School of Medicine; and a faculty member at the Johns Hopkins Public Health Preparedness Programs and The Johns Hopkins Bloomberg School of Public Health.

Suzann B. Goldstein is a medical sociologist and a freelance writer, poet, and author. In 2010, Suzann was elected to the Board of Trustees of the Foundation of the University of Medicine and Dentistry of New Jersey. In 2009, the Breast Cancer Center at the Cancer Institute of New Jersey, in collaboration with Suzann and her husband, Ed Goldstein, was renamed the Stacy Goldstein Breast Cancer Center in memory of their daughter, Stacy. In 1991, Suzann was awarded the New Jersey State Department of Health's Albert Harrison Award for Extraordinary Commitment to Services for Children with Special Health Needs. In 1976, in memory of their daughter Valerie, Suzann and Ed cofounded the Valerie Fund, an organization that now supports seven comprehensive health care centers for children with cancer and blood disorders in New Jersey and New York City.

Tegan Johnson is a clinical psychology PhD candidate studying at the University of Tasmania, Australia. Her research project focuses on the development of clinical psychologists, and she has interests in professional development, stress, challenge and trauma, positive psychology, and psychological growth.

Associate Professor David Johnston is the Director of the Joint Centre for Disaster Research in the School of Psychology at Massey University, New Zealand. The Centre is a joint venture between Massey University and GNS Science. His research has developed as part of a multidisciplinary theoretical and applied research program, involving the collaboration of physical and social scientists from several organizations and countries. This is one of the few truly multidisciplinary research clusters in hazard management internationally and is unique in New Zealand. He has been involved in developing integrated risk management strategies for different hazard events using techniques such as scenario development, mitigation planning, and community education programmes. The research has received numerous awards, both in New Zealand and internationally. David is a member of the Scientific Committee for the Joint International Council for Science (ICSU) and the International Social Science Council (ISSC) Integrated Research on Disaster Risk (IRDR); Royal Society Social Science Advisory Panel, on the Editorial Board of the *Australasian Journal of Disaster and Trauma Studies;* and Deputy Editor of *International Journal.*

Anne Links is a student at Towson University in Towson, Maryland.

John McClure is Professor in Psychology at Victoria University of Wellington. He completed his PhD at the University of Oxford, and his book, *Explanations Accounts and Illusions,* has just been published in paperback form by Cambridge University Press. He has published more than 60 peer-reviewed research papers, many of which focus on psychological factors that affect preparation for hazards, especially earthquakes. He led research funded by the New Zealand Earthquake Commission on factors affecting different types of preparedness in businesses and households and is researching judgments about low-frequency hazards such as earthquakes. He is currently doing research on the effects of the recent earthquake in Canterbury, New Zealand, on risk perception and preparedness.

Kimberley Norris is a practicing clinical psychologist specializing in adult mental health. She is an Associate Lecturer at the University of Tasmania. Her research interests focus on human adaptation and resilience with a particular focus on human performance in extreme environments. Her dissertation examined psychological adaptation of Antarctic expeditioners and their partners to Antarctic employment.

Douglas Paton is a Professor in the School of Psychology, University of Tasmania, Launceston, Australia. His research focuses on developing and testing models of resilience in high risk professions (e.g., emergency and protective services).

John M. Violanti is an Associate Research Professor in the Department of Social and Preventive Medicine (SPM), School of Public Health and Health Professions at the State University of New York at Buffalo, NY, and has been associated with SPM for 22 years. Dr. Violanti is also a member of the SUNY medical school graduate faculty. He is a police veteran, serving with the New York State Police for 23 years as a trooper, criminal investigator, and later as a coordinator for the Psychological Assistance Program (EAP) for the State Police. He has been involved in the design, implementation, and analysis of numerous police stress, trauma, suicide, and health studies over the past 20 years. His most recent study completion involved a 5-year study on psychological stress and cardiovascular disease outcomes in police officers. Dr. Violanti has authored more than 50 peer-reviewed articles on police stress and PTSD, police mortality, and suicide. He has also written and edited 15 books on topics of police stress, psychological trauma, and suicide. He has been an invited lecturer on topics of police stress and suicide to the FBI Academy at Quantico, Virginia, several times. He has lectured nationally and internationally at academic institutions and police agencies on matters of suicide, stress, and trauma at work.

Mary Beth Walsh is the mother of a young adolescent significantly affected by autism and a typically developing older teen. Although trained as a theologian, her advocacy work for individuals with autism reaches from faith communities to science-based interventions. She currently serves as Co-Chair of the Autism Task Force of the National Catholic Partnership on Disabilities and also serves as the Consumer Representative on the Board of the New Jersey Association for Behavior Analysis (NJABA). She teaches graduate courses in pastoral ministry at Caldwell College and is coeditor of the resource, *Journey into Community: Including Individuals with Autism in Faith Communities.*

PREFACE

The impact of events such as the 9/11 terrorist attacks and Hurricane Katrina were felt across the spectrum of organizations. Such events provide vivid illustrations of the exceptional circumstances that emergency and protective service agencies and businesses alike can encounter. These events stretched capabilities to the breaking point and sometimes beyond. Agencies and businesses encountered and had to cope with and adapt problems on a scale that far exceeded anything that their "routine" experience would have prepared them for.

If they are to respond effectively, agencies and businesses need to develop their capacity to adapt to unpredictable and challenging circumstances. To do so, agencies, institutions, and businesses and their officers and employees must be resilient. The subject of this book is how this outcome can be facilitated.

It is also a goal of this book to broaden the perspectives on the populations that need to be included when thinking about high risk groups and from whom insights into resilience and how it is enacted can be sought. Caregivers and groups existing or working in isolated conditions are considered.

The past few years have witnessed considerable growth in research into adaptation and growth outcomes in high risk professions. However, if organizations are to benefit from this knowledge and a strong foundation for future research put down, a resource that systematically integrates this work and highlights its significant implications and how they can be used in organizations is required. The proposed book will do this.

–Douglas Paton and John M. Violanti

CONTENTS

WORKING IN
HIGH RISK ENVIRONMENTS

Chapter 1

HIGH RISK ENVIRONMENTS, SUSTAINED RESILIENCE, AND STRESS RISK MANAGEMENT

Douglas Paton and John M. Violanti

How overcome this dire calamity,
What reinforcement we may gain from hope,
If not, what resolution from despair?
John Milton, *Paradise Lost,*
Book I, lines 189–191

INTRODUCTION

For many emergency and helping professionals, dire calamity is a fact of working life. Milton's quote challenges us to consider the fact that people have within their grasp the potential to influence the consequences of their exposure to dire calamity. The potential to draw on hope in changing the consequences of experiencing calamity foreshadowed the more recent recognition of how strengths such as hope and optimism can promote well-being and adversarial growth in the face of adversity (Joseph & Linley, 2005; Seligman & Csikszentmihalyi, 2000). Milton's quote also points to the role that resolution plays as a precursor to learning from the experience of dire calamity.

The *Oxford English Dictionary* variously defines *resolution* as "separating into components," "converting into other forms," and "formulating intentions for virtuous intent." The sentiments embodied in these definitions are echoed in the positive psychology and resilience

3

literatures with regard to securing salutogenic outcomes when faced with adversity. From the perspective of this book, the challenge posed by Milton's quote is to identify the beliefs (such as hope) and competencies (such as resolution) that can be pressed into service to turn the experience of calamity and despair into salutogenic outcomes in the form of resilience, adaptive capacity, and adversarial growth. The differences between the terms *resilience, adaptive capacity,* and *adversarial growth,* and the importance of distinguishing between them, is discussed in Chapter 9. But how are they related to risk?

RISK AND RESILIENCE

The search stimulated by taking up the challenge embodied in turning the essence of Milton's words into practical reality focuses on exploring two basic issues. The first concerns identifying what is meant by the risk posed by the dire calamites of everyday working and professional life. The second is about identifying how beliefs and competencies (e.g., hope and resolution) can ameliorate despair and facilitate a capacity to adapt in those for whom experience of dire calamity (e.g., critical incidents) is a fact of everyday life. In the contemporary traumatic stress and positive psychology literatures, characteristics such as hope and resolution are described by terms such as *strengths* or *resilience.* The title of this book alludes to their being a relationship between resilience and risk. The reason that a relationship between them can be discerned derives from how risk is defined.

Risk is the product of the likelihood of experiencing a challenging event and the consequences the event has for those who experience it (Hood & Jones, 1996). This relationship is illustrated in Figure 1.1. The events that are the subject of the contents of this book are those capable of causing physical, social or psychological harm to those caught up in them (e.g., hazardous events like natural disasters, acts of terrorism and mass transportation accidents).

According to Figure 1.1, risk is first influenced by the likelihood or probability of the occurrence of dire calamity. In the main, the populations (e.g., helpers, helping professions, emergency, and protective services professions) whose experiences are the focus of this text cannot control the occurrence of potentially hazardous events. Emergency and helping professions cannot readily influence the frequency,

Figure 1.1. The relationship between risk and resilience.

timing, nature, duration, or location of the events they are called on to respond to. However, emergency and helping professions agencies can make choices about the consequences their members experience as a result of experiencing a hazardous or critical incident (Paton, Violanti, Burke, & Gherke, 2009).

To facilitate this process, this book first aims to build understanding of how the "consequences" component of the risk equation emerges from interaction between environmental demands and the personal, social, and organizational resources brought to bear to deal with these demands (Figure 1.1). Next, it seeks to articulate how these resources can be applied (e.g., through selection, training, support, and organizational change) to ensure that the experience of calamity and despair is characterized by salutary outcomes and the development of the future capacity of people and agencies.

Consequences of Experiencing Critical Incidents

The consequences of critical incident exposure result from the potential of the characteristics of hazardous events to cause harm to populations. However, experiencing harm is not inevitable. The implications of exposure to hazardous circumstances can be moderated by the presence of factors that increase susceptibility to experiencing loss from exposure to a hazard (i.e., that increase vulnerability) and those

that facilitate a capacity to cope and adapt or adjust (i.e., increase resilience and adaptive capacity) following exposure (Figure 1.1). Because many of the factors that influence vulnerability and resilience
are amenable to change, this affords opportunities to manage risk.

Although typically associated with loss, risk also encompasses the
concepts of choice, anticipation, and resilience (Dake, 1992; Hood &
Jones, 1996). Dake pointed out that the contemporary tendency to
conflate risk with loss is a relatively recent phenomenon. The term *risk*
originally meant accounting for the gains and losses in games of
chance. Returning to the original and more comprehensive conceptualization, the risk concept becomes a future-oriented concept and one
capable of accommodating perspectives that cover adaptive and growth
outcomes and not just those associated with loss and distress outcomes.

More important, from the perspective of the present book, the
more comprehensive conceptualization of risk implies an ability to
make choices (through, for example, selection, training, and support
practices) regarding how people and agencies experience and interact
with threatening events. In particular, it points to the potential for
agencies to influence the likelihood of those caught up in such circumstances experiencing resilient and adaptive outcomes.

This more comprehensive definition of risk thus represents a suit-
able foundation for traumatic stress risk management in professions
and groups that face threatening and challenging circumstances repeatedly in the course of exercising their professional roles and for
whom the goal of risk management is to create the potential for the
experience of salutary outcomes. Risk management policies can then
be developed and appropriate practices and strategies implemented to
facilitate salutary outcomes as well as minimize the risk of pathological outcomes.

The foundation on which the attainment of these objectives is based
is risk analysis. A key element of risk analysis is the risk assessment
processes that identify the hazardous conditions that people may have
to contend with. Identifying the hazards, threats, and challenges that
people may face is an essential prerequisite to identifying the personal and collective resources and competencies required to be resilient
in the face of adversity. In this sense, the first step in the process of
understanding sustained resilience is identifying the threatening and

challenging circumstances that people and agencies have to contend with. In this context, it becomes pertinent to ask what is meant by a high risk group.

HIGH RISK GROUPS AND PROFESSIONS

Mention of "high risk" typically calls to mind the usual suspects: professions such as law enforcement, fire fighting, and the military. There is no denying that members of these professions do work in environments that regularly expose their members to challenging and threatening demands (Paton, Violanti, Dunning, & Smith, 2004). The high risk appellation is justified on the grounds of the overt and regular nature of their exposure to hazards, threats, and challenges on a daily basis.

However, other groups and professions, while not facing risk in such high profile ways, can nonetheless face significant challenges and thus merit being considered as high risk, at least with regard to some facets of their working life (Paton et al., 2009). This book thus discusses risk and resilience not only from the perspective of the "usual suspects," but also with regard to some less extensively studied groups that nonetheless face threats and challenges with sufficient regularity to warrant their being described as high risk.

In following this line, the objective is not to present a comprehensive overview of risk and resilience. Rather it is to introduce the need to cast the web of inquiry into risk and resilience in the work-related contexts wider. It is only by doing so that it will be possible to develop comprehensive theories of stress risk management. This, in turn, will provide the evidence base necessary to facilitate the development of more comprehensive practical strategies.

The adoption of the more comprehensive definition of risk, one that encapsulates experiencing gains and losses, has other implications for those groups and professions labelled as high risk. In particular, it implies that high risk groups have a greater propensity for experiencing salutary outcomes such as resilience, adaptive capacity, and adversarial growth. To realize the benefits implied in this use of risk, it is pertinent to consider how such outcomes can be attained. Identifying how this can be done is a challenge that has been picked up by the contributors to this volume.

Chapter Content

The contents of this book can be broadly split into two categories. Chapters 2 through 5 introduce the experiences of groups, such as mothers and Antarctic expeditioners, into the risk and resilience literature. The remaining chapters focus on the usual suspects, the protective, emergency, and military professionals who deal with threat and challenge on a daily basis.

In Chapter 2, Castellano, Goldstein, and Walsh illustrate the value of including hitherto ignored populations into the work-related risk and resilience literature. They aptly illustrate this in their insightful discussion of resilient mothers of special needs children. Their description of these mothers as "family first responders" captures the goal of this book as a voyage of discovery that seeks to broaden the perspectives on the populations that need to be included when thinking about high risk groups and from whom insights into resilience and how it is enacted can be sought. This is followed in Chapter 3 by Norris, Paton, and Ayton's review of resilience in Antarctic expeditioners, another group that has received little attention and can provide unique insights into how resilience can be sustained over prolonged periods of time.

In Chapter 4, Paton and McClure explore resilience from a business continuity perspective. The first part in the book concludes in Chapter 5 with Doyle and Johnston's discussion of resilience from the perspective of addressing how emergency managers can adjust to challenging operational circumstances by developing the capabilities to access and use scientific information to manage risk in complex emergencies and disasters.

These opening chapters provide insights into the lives, experiences, and outcomes of populations not usually considered as representatives of high risk groups or professions. The insightful and detailed observations recorded in these chapters illustrate why it is pertinent to do so.

The importance of these contributions can also be discerned in relation to their introducing a need for risk and resilience to be conceptualized as an ecological phenomenon. That is, one whose comprehensive appreciation can only emerge when resilience to dire calamity is conceptualized in terms of the interdependencies that exist among environments, people, professions, and agencies.

The second part of the book comprises chapters that focus on risk and resilience in the "usual suspects." For protective and emergency services officers and those on active military service, daily life can present challenges that could conceivably range from dealing with the carnage of a mass transportation emergency involving a freeway pile-up or a plane crash to mass shootings to dealing with the consequences of terrorist attacks using biological or chemical weapons. The uncertainty that members of these professions face with regard to their being unable to predict what, when, or where challenging circumstances will present makes developing resilience and adaptive capacity of paramount importance for these professions.

Furthermore, while the nature of police, fire, and paramedic involvement in critical incidents is often of finite duration, the emergence of terrorism and the growing need to be able to respond to prolonged natural and technological disasters (with events in Japan in early 2011 providing good examples of the kinds of circumstances that could be anticipated) highlights the importance of understanding how to develop resilience and adaptive capacity in the context of extended exposure to challenging and threatening circumstances (Paton & Violanti, 2007). Members of protective and emergency services professions can expect to experience this exposure repeatedly over the course of their careers. For those on active military service, prolonged exposure to threat becomes the norm.

The second part of the book opens in Chapter 6 with Drylie's account of a shooting from the perspective of the officer. While the traumatic stress literature frequently identifies events such as this as a traumatic stressor, it rarely presents the experience in the way it happened. Drylie's poignant depiction provides insights into the experience of a shooting in a way that puts resilience and adaptation into sharp relief. In Chapter 7, Everly and Links develop our understanding of the processes and competencies that contribute to resilience in their qualitative analysis of risk and resilience in law enforcement and elite military personnel.

Everly and Link's analysis introduces the role that leadership plays in developing and sustaining resilience. This issue is picked up in Chapter 8, in which Bartone and Barry discuss how leader actions and policies can, by creating a culture and context that increases mental hardiness, enhance human health and performance while preventing many stress-related problems before they occur. Bartone and Barry's work brought attention to the organizational context in which officers work.

In Chapter 9, in the context of drawing a distinction between resilience and adaptive capacity, Paton, Violanti, Norris, and Johnson expand on this and discuss the development of an ecological perspective that incorporates person, organizational, and family characteristics and dynamics in a comprehensive model of risk and resilience. In Chapter 10, the final chapter, Violanti continues this theme and, in the context of his Resiliency Integration Model, discusses how person and organizational factors can be mobilized to assess stress risk and to guide the development of effective stress risk management programs.

Collectively, the contributors to this book do three things. First, they bring a range of perspectives to bear on how high risk environments can be defined. Second, they identify the diverse ways in which sustained resilience is defined and enacted in a range of groups and professions. Third, they identify the strengths, beliefs, resources, competencies, and environmental factors that can be pressed into service in the process of developing and implementing stress risk management programs and strategies. That is, and paraphrasing Milton, the contributions to this book provide excellent examples of how people, groups, and agencies may be reinforced by hope and how they can gain resolution from despair in ways that increase the likelihood of their experiencing sustained resilience, adaptive capacity, and adversarial growth.

REFERENCES

Dake, K. (1992). Myths of nature and the public. *Journal of Social Issues, 48,* 21–38.

Hood, C., & Jones, D. K. C. (1996). *Accident and design: Contemporary debates in risk management.* London: UCL Press.

Joseph, S., & Linley, A. P. (2005). Positive adjustment to threatening events: An organismic valuing theory of growth through adversity. *Review of General Psychology, 9,* 262–280.

Paton, D., & Violanti, J. M. (2007). Terrorism stress risk assessment and management. In B. Bonger, L. Beutler, & P. Zimbardo (Eds.), *Psychology of terrorism.* San Francisco: Oxford University Press.

Paton, D., Violanti, J. M., Burke, K., & Gherke, A. (2009). *Traumatic stress in police officers: A career length assessment from recruitment to retirement.* Springfield, IL, Charles C Thomas.

Paton, D., Violanti, J., Dunning, C., & Smith, L. M. (2004). *Managing traumatic stress risk: A proactive approach.* Springfield, IL, Charles C Thomas.

Seligman, M. E. P., & Csikszentmihalyi, M. (2000). Positive psychology: An introduction. *American Psychologist, 55,* 5–14.

Chapter 2

FAMILY FIRST RESPONDERS: RESILIENT MOTHERS OF SPECIAL NEEDS CHILDREN

CHERIE CASTELLANO, SUZANN B. GOLDSTEIN, AND MARYBETH WALSH

INTRODUCTION

F *amily first responders* is a term used to depict mothers of special needs children who are living and breathing in a high risk environment. Crisis response, trauma, uncertainty, stress, and grief are among the experiences of traditional first responders. Similarly, in caring for a child with special needs, traumatic exposure and critical incident stress are also part of a family first responder's everyday experiences. How do we define family first responders' resiliency?

A brief search for a working definition of *resilience* found the following: the ability to recover from or adjust easily to misfortune or change (www.Merriam-Webster.com/dictionary, 2010); the ability to bounce or spring back into shape; the power or ability to return to the original form, position, and so on after being bent, compressed, or stretched; and elasticity (Webster's Encyclopedic Unabridged Dictionary of the English Language, 1996).

This chapter gives voice to three family first responders who epitomize each of these definitions. We share our stories in an attempt to define our resilience as caregivers.

Cherie's Story: "I'm Here"

The definition of resilience that I relate to most is "the ability to recover from or adjust easily to misfortune or change" (www.Merriam-Webster.com/dictionary, 2010). Adjusting to misfortune highlights that resilience is most needed in times of crisis, and the ability to remain fully "present" in these experiences is a key element for me.

In crisis counseling, there is a term called a *ministry of presence,* which refers to the moments when a person is fully present to another in a healing manner. Resilience can be a reciprocal process often found for me in moments of presence that are rich with emotion and risk.

When written in Chinese, the word *crisis* is described with two Chinese characters. They may be translated as *danger* and *possibility,* leading some people to suggest that a crisis represents a "dangerous opportunity." The most fundamental challenge for those who exhibit resilience is the ability to recognize the true nature of a crisis and the hidden potential within it. The resilient person not only minimizes the detrimental aspects of a crisis but also uses the situation to foster growth (Everly, 2009).

My path to becoming a resilient caregiver began at an early age continuing throughout my life experiences. I accept no credit for this journey because I believe it was divinely inspired. Considering my experiences, many family stories reflect my history of resilience in the face of misfortune. Reflecting, I have connected the past to the present to "map" an integration of these events that have paved my way and allowed me to be present despite the challenge.

Known jokingly as the "baton story," I entered a talent show in third grade. I dropped my baton at the beginning of the routine and it rolled off the stage! Rather than be mortified, I remained on stage and moved my hand as though it was still there. Dancing and fake twirling I actually faked a baton toss, and caught it of course! I was optimistic hoping the audience would think I twirled so fast that I had defied the speed of light! I lost the contest but believed that I would do better next time. The "baton" crisis was an opportunity to focus on my decision to remain on stage as a way of coping. I entered the contest confidently the following year and won.

A few years later, my parents brought me to the "Ice Capades." I had a chance to get in the show as a volunteer as the children were going to be in a parade float on the ice. My self-doubt and fear, however, derailed me, and I knew instantly that I should have simply tried rather than experience my first taste of regret. I literally "let the parade pass me by" with great dismay. My parents recall I screeched, "I will never let a chance pass me by again!" Another childhood memory that motivated me to seize opportunity and focus on persistence. In my pursuit of resilience, I have often required courageous initiative and persistence to cope with misfortune.

Beyond batons and ice skates, a real crisis occurred when my grandfather whom I loved suffered from manic depression. "Sunday macaroni" was delivered by my father and me to a psychiatric hospital where he was institutionalized before he died. Known as "Grandpa Moody" endearingly, this challenge prompted my empathy and fascination with the courage of people faced with mental illness. My role as a caregiver was a natural progression as I have found my sense of resilience in service to others. Being present in crises as opportunities for sanctified and intimate experience became a habit in my life. Resilience has been a learned trait for me with a conscious effort placed on action as a caregiver in service.

Persistent in my journey, my career included "crisis as a dangerous opportunity" as a daily occurrence at my job. Counseling in programs for traumatized first responders, police officers, military personnel, and mothers of special needs children, I am blessed with a capacity to be "present" with them. Peer-based helplines and crisis support services have offered shared resilience and lived experience fostering hope in times of despair.

Working in the aftermath of the events of 9/11, it seemed that resilience may also be derived from those around us. Hundreds of rescuers working on the pile, my husband among them as a volunteer, demonstrated awe-inspiring resilience in the face of an unimaginable horror. Heroic resilience uncovered in service.

Helen Keller stated, "Although the world is full of suffering, it is also full of the overcoming of it," and that has been true in moments where I have witnessed resilience and been inspired. My older son L. J. has often purely loved my younger son Domenick through difficult moments in overcoming his disability developmentally and physically. At the tender age of 11, L. J.'s resilience is clear in his acceptance and quiet calm. The purity of his love and vision guides his strength, and the summary of this challenge for him is simple. He explains, "My brother is special and he is perfectly Domenick." Studies of siblings of special needs children indicate that they may be more likely to become caregivers as a result of their experiences Perhaps L. J.'s road to resilience will lead him to become the light in the darkness that misfortune often brings. 9/11 hero's, my husband, my sons, patients at work, have all offered an opportunity to try to "catch" their spirit rather than lose hope. The more we are around each other in crisis, the more resilient we become. A simple truth in resilience for me is that there is strength in numbers. We are contagious.

The greatest challenge of my resilience was when my youngest son Domenick was diagnosed with special needs that warranted special doctors, therapies, and schools. We went to a famous pediatric neurologist in the area of developmental delays to provide him with his 238th diagnostic evaluation. (I'm joking; it's a few less than that.) Becoming professional parents with a special needs child, we packed away a toy chest worth of equipment, snacks from every major food group, multiple changes of clothes, and beverages so one might see us and think we were relocating or going on a vacation that required a passport.

Sitting in the exam room, Domenick decided he was going to make this doctor earn his money, and the process was difficult and painful both emotionally and even a little physically for our angelic child. As the session ended, the doctor sternly looked at my husband and I to answer our wrinkled laundry list of questions. My husband asked about our son's future, and the doctor answered with a story. He told us of a woman with a severe disability that now had a PhD. We smiled of course and nodded, but then he added, "She explained to me she doesn't even understand the concept of love, let alone feel the emotion." He basically inferred that our son would not experience love throughout his life, but perhaps we could teach him to paint or draw.

Our jaws dropped, and we left with our heads down, son in tow, and rushed out of the office and silently headed home. Our family was forever changed, and we were broken hearted.

Devastated, we tried to talk, ranting about how this famous doctor had gone senile and what he said could never be true. Exhausted a few nights later, I rested my head in my arms on the kitchen table while my family was eating some cheesy unhealthy sort of mush I made when suddenly I felt a hand on my head patting me gently. Too overwhelmed to look up, I waited and then the "music to my ears" was my son's voice whispering, "I wuv you" as his chubby fingers caressed my hair.

My husband's eyes met mine as he cheered out loud, my older son giggled, and Domenick's voice became the result of all the chances I ever took. "I will never let a chance pass me by again!" and I never have. I decided to take a chance on hope. This moment was a crisis as an opportunity for hope. A chance we could recover from this misfortune and become a family forever changed to become stronger and perhaps even more resilient.

Moms and dads with kids with special needs can often be in denial and avoid words to describe their child's problems. Some embrace the disability "martyr" role because they are so grief stricken by the victim role that they overexaggerate the severity of it. Others simply advocate often to feel some level of control. I have been all of these things depending on the day as now I see myself as something more than all the other opportunities in my life. I am a family first responder," I am a resilient caregiver. I am a mother of a child with special needs. Twenty four hours a day, seven days a week, I am a mother of a special needs child hoping I will morph into some heroine "Super mom" who could "Maintain all service needs in a single bound!" Often this crisis did not feel like an opportunity. It felt like a trial of faith, a challenge to be "present."

When all else fails, I find resilience in my faith. So I pursued spiritual guidance to forgive God and have strength to be grateful that I had been so blessed from my parish priest. I questioned our situation, and he responded by telling me, with his head tilted with a combination of what looked like pity and indigestion, "God knows you are the perfect person to cope with this. You are going to make a difference with this child and probably help others so it will all be ok." "What?" I volunteered at church and tried to be a decent person, so God chose me to handle this? Why didn't he choose my nasty ex-college roommate who treated everyone terribly or some mass murderer who deserves to be tortured? Not me! It did not seem fair, and my anger and self-pity engulfed my soul.

Searching for resilience to sustain I continued my struggle for spiritual guidance. I watched the movie "The Passion" several times but this one night I watched I saw something that changed my life forever. Mary sees Jesus carrying the cross through the crowd and he falls. Suddenly she has a flashback to when Jesus was a young boy and he falls while playing and scrapes his knee. She runs to him in her vision and recalls rocking him while he cried saying to him "I'm here, I'm here." The flashback ends and now she runs to Jesus through the crowd, holds him in her arms again and says "I'm here, I'm here." He looks at her and answers "Mother I will make all things new again" and he gets up and continues to carry the cross.

When my son cried and was sick but could not verbalize what hurt or what was wrong, I always comforted him, confidently announcing, "I'm here, I'm here." Those simple words say so much; it is an instinct, a motherly element in us, "I'm here." Instinct as a caregiver of a special needs child, or for anyone in need, is pure, so "I'm here" is a powerful phenomenon.

Suddenly I had the realization that Mary could be my model for this motherhood I had no idea how to rise to, and it became my focus.

A crisis now as a dangerous opportunity."I'm here" through the minor miracles in my life like my son's murmuring "I wuv you" just when I felt I could not go on. Soon after that experience, I began a faith-based support group with a nun friend titled "And a Child Shall Lead Us" for mothers with special needs children. After six years, the group has flourished and been replicated in other parishes. Perhaps that priest knew more than I did about the answers to my prayers. Resilience requires faith for me to flourish and truly be present as a caregiver of a special needs child.

Pondering resilience, I have found myself wanting to write happy endings as a family first responder. It has been as though I was cheering myself on to the next challenge. Sort of like training for an emotional marathon. The crisis and the ripple effects have an ebb and flow unlike any I've seen. In my professional life, I hear and witness suffering and the worst of our world. As a family first responder, I often see the best in our world, but I am overwhelmed by the stories. There never seems to be a beginning or an end. There are a million moments of beginnings and endings, births and deaths, of dreams, hopes, and then experiences that surpass expectations. Ebb and flow, and we try to normalize it by pretending that everything is OK.

I have often felt like when you have a moment when you are overwhelmed with grief hysterically crying and someone says something funny and in an absurd way you are laughing and crying at the same time? That's me. That messy, nose blowing, confusing picture of the gamut of emotions all combined into one moment. My experiences in this caregiver role have been layered in joy and pain, prejudice and warmth, and anger and faith like lasagna from heaven and hell! It is best described as a food analogy for me as an Italian American!

Domenick is doing beautifully, so we are one of the lucky families, but now forever guilty. Nose guard on a championship football team, mainstreamed with academic success, martial artist, wrestler, and singer, our son most significantly is described as loving! It is miraculous. Still we have wondered whether anything we did or did not do gave our child this issue. Guilty with the other special needs kids families because he has progressed in amazing ways we count our blessings. When their children are limited, we are almost embarrassed by his success, wishing we all could celebrate moving forward together, but we can't.

My caregiver resilience has been found in my presence in crisis as dangerous opportunities. Misfortune and recovery foster hidden potential found in myself and those around me with contagious resiliency. Ultimately, it is both my presence and acknowledgment of my weakness as part of my experience as a caregiver that has allowed my resilience to grow. Family first responders of special needs children require special resilience to avoid a bitter, faithless, angry existence. Being present through it all, I find opportunities for optimism, persistence, service, hope, and faith. Dangerous opportunities to be forever changed for the better. I'm not sure I have what it takes, but "I'm here."

Sue's Story: "Bounce"

Resilience can be defined as *the ability to bounce or spring back into shape.* (Webster's New World College Dictionary. 2010. Wiley Publishing, Inc.)

The prior definition of resilience applies to me with one qualification. I bounce but not as high as in earlier times. In that way, I see myself as analogous to a tennis ball that has been used too frequently. The fluffy felt has worn off, and, if one looks closely, the ball is misshapen. It has been forced to change yet it still bounces. That ball is resilient, as am I.

My husband Ed and I had two children: Valerie, diagnosed at three with bone cancer and gone from us six years later in 1976, and Stacy diagnosed at 25 with breast cancer and gone from us 12 years after in 2001. I cannot describe, in any way, the persistent sense of loss and pain we both feel. Our girls are gone.

My family now consists of Ed, Stacy's son Jonah, and loving relatives and friends. They have all helped to push me forward toward a new and modified state of resilience. I am grateful. Nonetheless, at the end of each day, and through all the years of grieving, I wondered how I was able to place one foot in front of the other, to keep on walking despite my losses. What was it that allowed me in all my various roles–wife, mother, caregiver, friend, advocate, back-to-schooler–to bounce upward from my down-facing position.

Although understanding came late in life, I have always believed that my past connects me to the present, not only through genetics but through family stories and personal memories as well. Surely, it is a built-in part of me. And so, I reason, those family stories and my memories are major contributors to my resilience. Acknowledging that resilience, my bounce, has also played its necessary part. For years, people have told me that I was strong. I'd hear, "Sue, I could never do what you do" or "Sue, I could never have handled it." Or inevitably, "Sue, you are so strong," And I would respond wordlessly, "What would you do, run away? Leave your loved ones to fend for themselves? What?"

Their comments angered me. I rejected the idea of my strength, my resilience, because I felt so completely destroyed inside. Yet, in time, I learned that it didn't matter how I felt or how I was perceived by others because I did what had to be done, and I did it with all my heart. The realization came hard, but I now recognize it as the truth. I am relieved. I am resilient, and I know it.

Time has passed and I have changed, as we all have. The memories of my loved ones, too, have changed, transitioning with the years from pure anguish to warm recall. Whether changed or not, they are with me always: that blank traumatic space filled with stories of love about my mother who died when I was not quite 10; my big brother Stan–my best friend–who died when he was 26 and left me with vivid memories that I still consider most precious; my fun-filled father, residing alongside Stan, who became the epitome of the loving, single parent when that term didn't exist; and my wondrously soft, surprisingly strong grandmother who was invariably at my side, or in the background, whenever one or the other was needed. I considered these latter three my lopsided life and blood family, each supporting me without fail as I grew up. But all four long-gone family members have helped to link me gently to the present.

That's not all. Of course not. I have left, for last, the most recent and most painful losses of all: our girls, Valerie and Stacy. Their memories include the ongoing disbelief that they are, indeed, gone, and the sustaining moments, the in-between times when life was lived with a big grin and a great bounce.

Stacy and Valerie stand at the forefront of my special collection of memories; they recall for me learned experiences layered one upon the other. Our children taught me to adjust to life's bad stuff and to pass that ability on. Talking about them with Ed and sharing their stories with others serves to further bolster my memories. Leading my long list of loved ones, their memories provide the means to my resilience, to my new and forward-looking existence.

Valerie's diagnosis in early 1970 pulled my little family of four into an unfamiliar and chaotic lifestyle filled with risky chemotherapy, radiation treatments, and oncology visits. But during the in-between times, the good times when she was away from the medical system, my little girl bounced high and often. Ed, Stacy, and I followed–Ed and me admittedly with much fright and less bounce. It was a different form of resilience, but, nevertheless, it was resilience.

For instance, I remember, during Valerie's hospitalizations several years into her illness, when parents were finally permitted to sleep over. I had decided, at that point, that the best way to handle Val's hospital stays was to leave my chair-as-bed early–earlier than the doctors making their rounds–run to the shower, dress, and return to Val's room looking as if I had everything under control.

Many of the other mothers on the floor appeared disheveled and distressed. They were frightened, as I was, but they also appeared frightened and, therefore, in my eyes, out of control. I thought, how did their sick children react? They were frightened of being ill and in the hospital, but additionally were they frightened by their mother's appearance? I wasn't sure about that, but I would not take the chance with my child.

I tried not to be judgmental about those moms, but I would not be like them. I believed if I looked in control, then I would be in control, for my daughter and for myself. What was inside would stay inside. We each benefitted from that effort. In the mornings, when my daughter awakened in her hospital room and saw me neatly dressed and ready for the day, she would smile and say, "Hi, Mommy." I would smile back and say, "Hi, Pussy Cat." Despite the circumstances, we bounced.

Val's bone cancer recurred in 1973, once again in the tibia of her right leg. The doctors said amputate. There were no other choices. Because the cancer was found nowhere else in her beautiful, little six-year-old body and fearful of metastatic disease, Ed and I agreed.

The night before Val's surgery, we made the unhappy arrangements for the next day. Ed would go home after spending the day with Valerie and me. He would tell Stacy that her sister's leg had been amputated above the knee. I'd sleep in Val's hospital room and take care of my little one when she was fully awake.

Two long hours after the end of surgery, Ed and I were finally allowed into Valerie's room. She lay sleeping, her bandaged right thigh resting on a large pillow, a blanket drawn up to her chin. We hovered over her, stroked her, kissed her, but heavily sedated, Valerie was, for the most part, unaware of our presence.

Ed went home to Stacy that night, and I sat in a chair next to Val's bed, waiting for the new day to begin so that I'd tell her—what? Exactly how would I handle her questions? What should I say to her? What would Valerie understand?

Early morning arrived. I stood up, straightened Val's blanket, and sat back down waiting for her to awaken. Whatever bounce I had was grounded on hope: that our decision had been the right one, that the outcome would be successful. Sitting quietly in the chair, my legs resting on the edge of Valerie's bed, I watched a bright sun rise up outside the hospital room's one dingy window. My thoughts, wandering far from that awful place, were interrupted when I heard, "Hi Mommy."

I stood up, leaned down, said "Hi, Pussy Cat," and gave my daughter's pale cheek a light kiss. "Look, Mommy," my cheery little kid said. "The doctor told me he was taking away my whole leg. But he didn't. He left me a little leg!" With Val's hand on my arm, that spirited child lifted high her bandaged thigh, her "little leg" and introduced me once more to her own, very magical slant on life. And I began to bounce upward at that most wondrous of ongoing surprises, a child's matchless view of life.

Although only six, her resilience was catchy. Looking back, whatever additional bounce I had was reinforced by the Valerie Fund, the organization that Ed and I founded in Val's memory for children with cancer and blood disorders. People were generous with their time and money. They helped pursue our idea that sick children and their families needed highly rated pediatric oncology care in hospital-based medical centers throughout New Jersey. The idea took hold and grew into eight children's centers across New Jersey and in New York City.

The Valerie Fund was our way to help children within our state, like Valerie, who were severely ill. A dynamic grassroots movement, the Valerie Fund reaffirmed for me the vigor that had characterized Valerie. Its multitude of functions allowed me to stand tall and walk with a purpose. Reaffirming, too, was the memory of one Valerie Fund mother. I don't remember her name, but her image is clear. Let me tell you about her.

I was walking down the hall of the Valerie Fund Children's Center at Overlook Hospital one day when I was suddenly pulled into a small examining room by a slight, middle-aged woman with short brown hair. Her son was making zooming noises with his throat while pushing a toy truck on the floor. She hugged me, then let go, but held tight to my hand. In front of her little boy, she said, "I wish you had a Valerie Fund Center to go to when your daughter was sick." We talked for a minute or two, and then I left, reeling from her enthusiasm about what Ed and I and our organization had done. On the way back to my car, I grinned, nodded my head, thought we did it, and couldn't stop the upward bounce. Didn't want to stop it.

More time passed. My good memories labored to override the bad. Until two years into Stacy's marriage. At the age of 25 our surviving daughter discovered a lump in her breast; a biopsy declared the lump malignant. I stood beside her bed after the procedure, and, though groggy from anesthesia, Stacy squeezed my hand and grinned wryly, as if to say, "Well, Ma, what did we expect?" and then with another squeeze, she said, "We'll deal." I tried hard to answer her smile with mine.

Though Valerie's confrontation with cancer had failed, Stacy faced her illness with courage, undertook treatment, and when it became too hard at any point, went to bed and pulled the covers over her head. Renewed by the warmth and quiet of her private space, Stacy did what had to be done; like her sister before her, she fought the illness that threatened her and bounced back during the in-between times. Ed and I followed along slowly but surely. Our resilience took effort.

One year after the diagnosis, mastectomy, and chemotherapy, Stacy's oncologists were unable to detect any signs of cancer. Our daughter resumed her life without pause. Soon after, Stacy's breast cancer recurred, and once more, we were offered a no-choice, drastic type of treatment: an autologous bone marrow transplant. Studies had not proven one way or another that it was beneficial for breast cancer, but we grabbed at it anyway.

Stacy and I spent the days together in her hospital room at Johns Hopkins Medical Center's transplantation unit. We talked when my daughter wanted to talk, and whenever she wasn't vomiting or sleeping, I'd follow her orders and pick up extra head scarves at the Inner Harbor, find Chinese food for lunch, and new books to read. I'd leave, reluctantly, after dinner to sleep at a nearby hotel.

The bone marrow transplant was a grueling process, but, at last, it was finished. Before Stacy was discharged, however, the doctor who ran the transplantation unit gave her a list of instructions for the next six months or until her blood counts had risen as close to normal as possible. His instructions included no food that was raw or smoked, no large groups of people, a face mask worn when outside, and as much rest as possible.

The day after leaving the hospital, Stacy, her husband Robert, Ed, and I piled into Robert's car and headed back to New Jersey. After dropping us off, the young couple drove on to their condo and back to life in East Hanover. I forced myself to wait until the following morning to phone and find out if all was well with her." Stacy, how are you, Honey? Did you sleep? Did you eat? Did you go right to bed when you got home?" "Oh, Mom, we went to the movies, a matinee. It was such fun." "What? You're not supposed to be near groups of people until your counts are up! Did you wear a mask?" "The theatre was empty, Mom. Don't worry. I know what to do. I'll do what's right. But I'll do it my way." And with those words, I came across, once more, Stacy's determination and her bounce. Of course she didn't wear a mask. Of course she was ready to resume her life. And with a nod of my head, she had reminded me to bounce along with her. No doubt, resilience was contagious.

Not too long ago, on one of my walks, I stopped as I usually do, in front of a small, cactus-like tree with spindly branches that reached up to the sunny blue Florida sky. The tree is mostly barren, but two or three times a year it yields a few extra-large white blossoms. They are gorgeous. The tree in bloom, or bereft of them, reminds me of a Georgia O'Keefe watercolor. I love it and call it Georgia.

On this particular day, standing in front of Georgia, my eyes filled with tears. I felt devastated. Why weren't my girls with me? They belong next to me witnessing the beauty of nature; we could talk about the tree, be in awe of its blossoms, and all three of us would call it Georgia. With tears in my eyes and without any warning, a vision of Valerie in her Minnie Mouse T-shirt popped into my head. She would have been the first to tire of the tree's beauty; pulling at my hand impatiently, she'd say, "Come on, Mommy, let's go." My tears dried. I began to smile.

And I am reminded that while my tennis ball remains misshapen, it still bounces. Now, instead of thinking why my girls aren't standing beside me looking at Georgia, I remember that my girls are part of me, that I have three pairs of eyes to see with: mine, Valerie's, and Stacy's.

My memories, those bits and pieces of my past, strengthen my world. They remind me, again and again, that my long-gone loved ones were resilient. As am I.

Marybeth's Story: "Why Not Me?"

Resilience defined as; the power or ability to return to the original form, position, etc., after being bent, compressed, or stretched; elasticity. (Webster's Encyclopedic Unabridged Dictionary of the English Language, 1996.

Bouncing back recalls for me those bowling pin-shaped blow-up toys from my childhood that were weighted at the bottom and painted with clown faces. You could punch them, kick them, and knock them down again and again, but they would always bounce back into the same position they started–upright and ready for the next blow, painted smile undaunted by the pounding they received.

Thinking of resilience as that clown image snapping back into its original upright position helps to show that resilience is a morally neutral designation. People who bounce back from challenges in their lives are not necessarily better people than those who do not fare as well. A whole idiosyncratic mix of things has contributed to making me resilient, and I am loath to take credit for almost any of them.

When my younger son, Ben, was diagnosed with autism, we threw ourselves, as many parents do, into helping him. Despite the sage advice to recognize that autism is not a sprint, it's a marathon, we quickly became one of those families with professionals in our house working with Ben for 50 to 60 hours a week. We took our school district to court to secure Ben an education from which he would benefit. We tried our best to make every waking moment a teaching opportunity.

Ben had to be taught almost everything: sitting in a chair, looking another in the eye, tolerating having his teeth brushed or hair cut, eating solid food, sleeping through the night, and using the bathroom on his own. This took enormous efforts over many years, but when you have a child with extraordinary needs, you don't really have a choice. I suspect that many caregivers feel this way, that we don't have a choice, and for that reason thinking of ourselves as resilient can be uncomfortable.

Yet those of us who have given care for many years, and who look at the future seeing no end in sight to caregiving, can benefit from thinking about what has sustained us and what precisely carries us through each day. It's just important to preface that reflection with the assertion that most of what has gone into making me resilient can be chalked up to either luck or accident–I seek no credit here. That partly explains why my favorite definition of *resilience* is the ability to bounce back.

I suspect that I am simply hard wired to be one of those people who wake up mostly happy–and this certainly contributes to resilience. Although I had watched loved ones wrestle with the demons of depression and wondered why and how it was that I never felt that way, it was my son's autism diagnosis that made me think more deeply about temperament.

After one has experienced a life-changing loss, be it a death, a shattering diagnosis, or other news of a tragedy that will affect you in a permanent way–that liminal time between sleep and wake looms large. After Ben was diagnosed, I would occasionally experience, as I was transitioning from sleep to awake, a moment of psychic peace, feeling again how I used to feel before his diagnosis, that all was right in the world, that I was lucky to be in my own warm bed, with my spouse who loved me beside me, that I was simply happy to be alive for another day. But in the early months postdiagnosis, when we were sad all the time and panicked much of it, this sense of peace and calm was especially fleeting. The second my brain was awake enough to recall Ben's diagnosis, to begin to remember where I was in time and space, in that very instant, the calm was shattered, and with it all my feelings of well-being drained away. I experienced this as vexing and was annoyed at myself for forgetting–even for one moment, even while sleeping–the terrible and daunting challenges that lay ahead for my young son and my small family.

It took time, but I finally figured out that waking happy, the fleeting mini-moment of contentedness I still experienced despite the changed reality I now inhabited, was quite beyond my control. It seemed to me that I was hard wired to wake this way–and nothing could change this. And I felt that my neural hard wiring, the incomprehensible way my brain was structured was wholly beyond my control–in exactly the same way that Ben's brain wiring, the very aspect of him that resulted in his autism diagnosis, was completely beyond my control.

Relishing the irony that the same incomprehensible brain wiring both led me to wake happy and led Ben to be diagnosed with autism somehow made both realities easier to accept. My insight that I may be temperamentally suited to resilience is an example of but one of many factors that has resulted in my actual resilience. Of critical importance for me, however, has been the ability to make sense of Ben's autism diagnosis.

Of course, that's an absurd claim. No one can make much sense of autism. Autism has no known cause and no known cure; scientists can only speculate about the autism mechanism. There's no blood test or high-tech scan that can diagnose autism, and no one can possibly tell me why my younger son has it while my older one does not. Autism remains deeply inscrutable–and like so many other realities of human life that make no sense in a standard, Western scientific understanding of the world, like the suffering of innocents from cancer, war, or natural disasters, sometimes it helps to turn to different explanatory systems. For some of us, this means turning to our faith.

Not that these are easy questions for people of faith. Remember Job, the just man of Hebrew Scripture, who suffered extraordinarily–and who railed at God demanding explanation, all in vein. But as a person of faith, and an academically trained theologian, I did find in my many years in seminary, a conversation about suffering and injustice and God that proved to be a lynchpin in my ability to be resilient when needed.

Personally, I needed to be able to make sense of, to understand as best I could, the reasons for Ben's autism, the place where our story sits within the context of our faith claims. Only when he was diagnosed did I come to realize that the theological cosmology I had adopted in my years in seminary could actually support and sustain me now.

I entered the study of theology propelled by a strong sense of outsider status as a woman in the Catholic Church. Early study was fueled by the outrage ignited at the treatment of women in the history of Christianity, the misogynistic lens through which texts and traditions had been read and distorted, the patriarchal power structures that continue to exclude. I was propelled to join the theological conversation, studying over a decade in seminary, by my strong identification with those who had been wronged, by those whose experience of injustice propelled them toward the God of justice with hard questions and wounded hearts.

This led me to study a specific sort of theology, that which arises out of the matrix of powerlessness and injustice and still manages to assimilate those searing experiences into the basic Christian narrative of a God who loves us all enough to become human in order to carry us back to God's own self in love. This type of theology is called Liberation Theology, which embraces many varied points of departure, from the experience of the poor in Latin America, to that of African Americans in the United States, to that of gay and lesbian people in the church and marginalized women as well.

Theologies of liberation take as their starting point the human experience of oppression or injustice, and they see in their own experience a reflection of precisely what God experienced in Jesus. From this vantage point, the Gospel tells the good news of a God who abides with humanity through the most brutal suffering, whose presence is to be found up to and through the moment of death, and who vindicates through love and beyond death. The Gospel story of Jesus' life, death, and resurrection underscores the phenomenological reality of suffering and the very humanness and inescapability of oppression and injustice.

But my long sojourn in seminaries taught me so much more than the theology I could absorb by reading books. I may have entered angered at the church's treatment of women through history, but what I encountered in seminary was friend after friend whose personal story of oppression and lived experience of injustice was so much more of a life and death struggle than mine that I could hardly believe some were still standing.

I learned much from these friends, from Black South African friends who had grown up under apartheid, from friends who had escaped the civil war in El Salvador only to leave their family and hearts behind, from courageous and scarred gay men and lesbian women who have wrestled for years with that still small voice calling them toward service in the church that was filled with loud and angry voices condemning them as sinful and guilty. By listening to their stories, by reading their words and the words of our teachers, something changed for me. In fact, two things changed decisively for me in my seminary years. First was the radical reorienting of my most cherished faith claims; liberation theology taught me how to hear the promise of the gospel in light of the very human experience of injustice—and to see that it still, more than ever, spoke of God's faithfulness and love. This was a tremendous gift. This libratory reading of Christian belief is what has allowed me to remain Catholic.

The second thing that changed for me was the slow unraveling of my rather unexamined sense of entitlement. When I first went to seminary, fueled by the injustice of women's treatment in the church, that anger was partially grounded in the recognition of actual injustice and partly based in my naive sense that I personally don't deserve this second-class citizen treatment. After all, don't I go to church each week? Don't I contribute? Am I not kind to small children and wounded animals? Don't I try to lead a life pleasing to God? Am I not a good person? Well, if so, then why am I experiencing injustice–it's not fair!

The cry, "It's not fair!" can be a real stumbling block to resilience. It is so, so easy to get stuck in the quicksand of self-pity and become bogged down by feelings of resentment. It is also temptingly easy to get lost in the cry of "It's not fair!" when all one has known is the comfortable existence of the average middle-class American. We Americans have practically forged our national identity by appeal to values like rugged individualism and self-determinism. Scratch the surface of these very American motifs, and you will find a set of unspoken assumptions that life is basically fair, that we are in control, and that good things happen to worthy, good people.

Much has been written about the influence of certain Christian views of predestination and double predestination in the creation of a distinctively American self-understanding, and I'll leave that to better scholars than myself to plumb more fully. But this I know–I was certainly motivated to study theology both by my conviction that the oppression of women in the name of Christianity was wrong, and also additionally–it was just not fair!

The second decisive thing I learned in seminary was that all those friends and teachers who had lived through much greater life and death struggles than I had evidenced very little sense of the "It's not fair!" cry.

Few of my friends and teachers were white, middle-class Americans. Almost none wrestled with feelings that the things that had happened to them were unfair or that they didn't deserve these hard and scarring encounters with injustice. My friends, products of communities shaped by injustice over generations, helped me to see that God does not mete out pleasantries and hardships in people's lives as signs of God's favor. Growing up African American in the south, growing up poor in San Salvador, growing up black under apartheid in South Africa–my friends and teachers came from families where generations had lived with injustice and had come to see it not as a judgment but rather as a fact of life.

Fundamentally wrong, to be sure, and professed in faith as an affront to God, but a fact of life nonetheless. Letting go of this perception was perhaps the most valuable thing I learned in seminary, but it came over time like sand slowly slipping through the fingers of a cupped hand, until eventually, I felt sure that I understood–the things that happen to us in our lives are not signs of God's pleasure or displeasure with us. Tragedy occurs, injustice abides, suffering is part and parcel of human life, what matters–to us and to God - is how we respond to the challenges we encounter. Years of reading liberation theology and listening to my friend's stories have taught me this.

Or so I thought. But then came that bright May morning, when all alone, I took my adorable but curiously unhappy little two-year-old to a developmental pediatrician, who confirmed for me that Ben had autism. I walked him back out to the car, strapped him in his car seat, got into the driver's seat, and just cried. I knew then with absolute certainty that our lives were from here forth going to be different. I knew that we were no longer just a typical family with two healthy, smart kids. I knew that this was a seismic change, this was permanent, and from now on people would look at us differently and with pity. None of this change was easy for me at first, but never for a second did I think, "It's not fair!" because I had long ago abandoned that type of thought. In fact, I called my Salvadorian friend from seminary, who was also Ben's godmother, and told her of Ben's diagnosis. And I confessed to her that although I had thought I understood all she had told me of her life and her family's struggles over the years, I really hadn't–but I understood better now.

Another autism mom who is a good friend once turned to me during a support group and said, "I don't understand why parents ask, 'Why me?'"

"I've always thought," she continued, "Why not me?"

So I know that some people understand the randomness of life, the tragic inequities, the awful fact that bad things do happen to good people, without spending many years working their way through seminary to get there. But that's how I needed to arrive at the insight. And for me to really be able to own this insight about the basic unfairness of much of life, I needed to be able to situate it within the larger confessional context that liberation theology provided me.

Ben is today almost a teenager. He is happy, healthy, and well loved. He will never drive a car, marry, or live on his own, but we are hard at work still to make sure that he will have a good life. There are many, many things that have helped me be resilient in caring for him over the years of which I have not spoken—our tremendous luck in securing quality intervention from almost the moment of diagnosis, our ability to fight to secure him an authentic education, the wide web of generous and supportive people who have sustained us—family, friends, and professionals. And most important also is the wide range of advocacy work I do, in professional, educational, and faith-based settings to promote the well-being of children with autism and other special needs well beyond my own son. This work is what sustains me now. But the preparation for this work occurred years before in those halcyon days of my graduate education in theology.

I still believe that my best preparation for this journey has been my seminary education. Years spent learning to let go of the innocent assumption of fairness, coupled with growing certainty in a narrative of God as the one who does not reward us with an easy life, but rather promises to be present in all of life's challenges, were for me the necessary prerequisites for resilience.

To summarize, these three stories have defined resilient caregivers in a new framework as *family first responders.* All three women have described their experience as being present, bouncing with love, and gaining insight into faith. Our stories confirm that we have been forever changed, redefined in resilience, cherishing what we now know as "life in between." "Life in between" describes the cherished time in which we are loving and living amid our special needs in the simple joys found in moments at a kitchen table, a hospital room, or a church confessional.

The clear message from our voices are themes such as our past and present, love, service, faith, opportunity, justice, and inspiration. In the final analysis, we are hoping for an epidemic contagion of family first responder resilience as it seems our voices are ready to be heard.

REFERENCES

Everly, G. S., Jr, & Strouse, D. A (2009). *The secrets of resilient leadership: When failure is not an option.* New York: DiaMedica.

Merriam-Webster Unabridged. (2010). *Merriam-Webster's Collegiate Dictionary* (11th ed.).

Webster's Encyclopedia Unabridged Dictionary of the English Language. (1996). New York: Gramercy Books/Random House.

Webster's New World College Dictionary. (2010). Cleveland, OH: Wiley.

Chapter 3

STAYING COOL UNDER PRESSURE: RESILIENCE IN ANTARCTIC EXPEDITIONERS

KIMBERLEY NORRIS, DOUGLAS PATON, AND JEFF AYTON

INTRODUCTION

Due to its freezing temperatures, fierce katabatic winds, and geographic isolation, Antarctica is widely cited as one of the most extreme and unusual environments on Earth (e.g., Suedfeld, 1991). The nature of this environment renders it impossible to sustain human life without the aid of technology and complex operations, and its geographic isolation limits the availability of external assistance in medical and other emergencies (Decamps & Rosnet, 2005; Lugg, 2005). Challenges associated with the physical environment, climate, and geographic isolation inherent within Antarctica increase the risk of injury or even death to those who reside there (Palinkas & Suedfeld, 2007). In addition to the physical risks and demands placed on those working in this environment, Antarctica also poses additional and unusual social and psychological challenges for expeditioners which need to be negotiated to enable successful adaptation to occur. These challenges include, but are not limited to, prolonged separation from family and friends (Norris, 2010).

Existing research on the psychological issues that affect returning Antarctic expeditioners testifies to the coexistence of both positive and negative outcomes derived from their experience "on the ice" (Palinkas, 2003; Taylor, 1973; Wood, Hysong, & Lugg, 2000). However, the predominant focus of such research has been limited to adjustment outcomes rather than the processes that underlie adaptation–processes that are likely to include individual coping mechanisms and family functioning patterns (Houtzager et al., 2004). Additionally, most research (including that conducted on Antarctic populations) has investigated vulnerability and resilience at the individual level, despite an increasing awareness that a comprehensive understanding of resilience must integrate individual, organizational, and relationship perspectives (Paton et al., 2008). This argument is based on the fact that both organizational and relationship environments define the context within which individuals experience and interpret events and within which future capabilities are nurtured or restricted (Paton, 2006). Therefore, expanding the research focus beyond the employee to include traditionally overlooked parties (i.e., partners) is essential to developing a comprehensive understanding of the separation experience within the context of the family, as well as individual and organizational units. Identification of the salient predictors of resilience and articulation of the mechanisms linking them to adaptive outcomes for expeditioners and partners alike will enable intervention strategies to focus on enhancing this capacity throughout the separation experience, thereby enhancing both performance and retention.

UNDERSTANDING RESILIENCE

Exposure to challenging events can create a sense of psychological disequilibrium that represents a situation in which the existing interpretive frameworks or schema that guide expectations and actions have lost their capacity to organize experience in meaningful and manageable ways (Janoff-Bulman, 1992; Paton, 1994). The challenge is thus to identify those factors that can be developed prior to exposure that reduce vulnerability toward experiencing negative outcomes and enhance an individual's capacity to develop schema that broaden the range of (unpredictable) experiences that can be rendered coherent, meaningful, and manageable (Frederickson, 2003; Paton, 1994, 2006). In this way, the likelihood of experiencing resilient outcomes is enhanced.

At present, the most comprehensive model of resilience within the literature is the Stress-Shield model (Paton et al., 2008). Developed to explain the processes contributing to resilience in police officers, this model proposes that resilience reflects the extent to which individuals and the groups to which they belong can capitalize on resources and competencies (both psychological and physical) in ways that allow challenging events to be rendered coherent, manageable, and meaningful (Paton et al., 2008). Furthermore, it acknowledges the contexts in which individuals experience and interpret events and their consequences and within which future capabilities are nurtured or restricted (Paton, 2006). In its original iteration, the Stress-Shield model did not include familial influences in the resilience and adaptation process. The model has recently been revised to include family (see Chapter 9). The importance of considering family is central to the argument being proposed in this chapter.

Previous researchers have consistently demonstrated that even if only one family member is directly exposed to the challenging situation, other members of the family unit are indirectly affected by the impact that changes in the individual's functioning have on preexisting family functioning and relationship dynamics (Stinnet & DeFrain, 1985; Walsh, 1996, 2003). In turn, these same researchers have demonstrated that family processes (including the presence of an intimate relationship and family-level coping strategies) mediate the degree of adaptation demonstrated in response to the challenge, thereby necessitating family/relationship-level factors be included in any comprehensive model of resilience and adaptation.

Furthermore, considering the unique nature of Antarctic employment and distinct differences from other forms of employment (including police work from which the above model was primarily derived, and which differs in terms of the duration, nature of experiences, and separations from family members), the degree to which existing models of resilience can be generalized to Antarctic populations has not been considered. Another unique aspect of Antarctic employment is the systematic, predictable, and repetitive aspects of exposure to an extreme environment within a professional capacity that can provide greater opportunity to develop proactive prevention and intervention strategies both prior and subsequent to exposure. In this vein, the importance of understanding the mechanisms underlying resilience within such populations is further emphasized.

For this reason, the current authors sought to investigate the processes underpinning resilience within a population of Australian Antarctic expeditioners and their partners. Employing a mixed-method, cross-lagged longitudinal design, it was possible to consider the concurrent influences of individual, organizational, and relationship-level factors in the development of resilient outcomes in expeditioners. Furthermore, by examining changes throughout the employment experience from predeparture through absence, reunion, and reintegration, it was possible to identify the mechanisms underpinning these outcomes.

Modelling Resilience within Antarctic Populations

The overall findings of this research demonstrated that the majority of expeditioner and partner responses identified few long-term negative effects associated with Antarctic employment experiences. In fact, there was evidence of improved functioning across individual, organizational, and relationship domains beyond that reported at predeparture consistent with not only resilience but also growth outcomes. These improvements were demonstrated not only in the immediate aftermath of the expeditioner's return from Antarctica but sustained 12 months postreturn.

It is noteworthy that prior to departure for Antarctica, expeditioners and, to a lesser extent, partners demonstrated high levels of optimism, personal growth initiative, quality of life, well-being, and the use of active coping strategies at both the individual and relationship levels (see Table 3.1)–traits associated with resilient outcomes in other populations (e.g., Patterson, 2002; Southwick, Vythilingam, & Charney, 2005).

Furthermore, these high levels of functioning were reported despite high workloads, which created conflict between work and family roles. Although the mean levels of each of these positive indices fluctuated over time in response to challenges encountered throughout the employment experience (primarily during the absence period for partners and the reunion period for expeditioners–see HSCL-21 scores presented as an example in Figure 3.1), at 12 months postreturn from employment, they had either returned to or improved on predeparture levels.

To understand the mechanisms facilitating these outcomes, functioning within and between stages of the employment experience was examined through the use of qualitative interviews, with surprisingly

Table 3.1
DESCRIPTIVE STATISTICS FOR INDICES OF RESILIENCE REPORTED BY
EXPEDITIONERS AND PARTNERS PRIOR TO DEPARTURE FOR ANTARCTICA.

Variable	Expeditioners		Partners	
	Mean	SD	Mean	SD
PGI (Max=54)	32.99a	6.67	27.13b	7.30
LOT-R (Max=24)	21.97d	4.12	22.14d	3.49
HSCL-21 (Max=84)	29.76f	6.54	33.10g	6.43
WHOQOL-BREF (Max=5)	4.66i	.48	4.58i	.50
COPE Active (Max=16)	11.76k	2.19	12.92l	1.89
FFSS Coping (Max=100)	73.54m	14.54	71.30m	14.60

Note: Means not sharing the same subscript are significantly different; PGI=Personal Growth Initiative; LOT-R=Life Orientation Test Revised; HSCL-21=Hopkins Symptom Checklist, 21 item–inversely related to well-being such that low scores indicate higher levels of well-being; WHOQOL-BREF=World Health Organisation Quality of Life Inventory–Brief; COPE Active=Active Coping measured by COPE; FFSS Coping=Family Functioning Style Scale Coping Dimension.

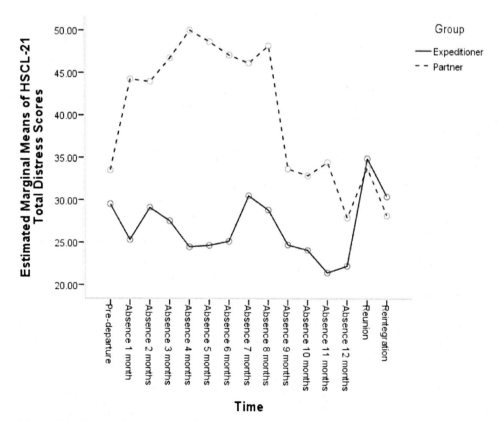

Figure 3.1. Fluctuations in mean distress reported by expeditioners and partners throughout the Antarctic Employment Experience.

consistent themes being identified as contributing to positive experiences. Specifically, it was identified that the primary task associated with each phase of the Antarctic employment experience (whether it be predeparture, absence, reunion, or reintegration) involves cognitive restructuring to accommodate new information, experiences, and expectations–that is, develop a new schema–thereby facilitating enhanced functioning and subjective well-being, which are indicative of resilient outcomes. Factors found to influence this cognitive restructuring related to the availability of adaptive psychological resources, a collaborative organizational climate, positive relationship dynamics, availability and satisfaction with social support and information exchange, perceptions of trustworthiness (in terms of personal competency, partner behavior, and organizational support), and a sense of empowerment.

Importantly, the degree to which this restructuring would occur related to the presence of these factors throughout the employment experience, not just 12 months postreturn from Antarctica. What this suggests is that by engaging strategies to proactively facilitate these factors prior to departure and throughout the employment experience (not solely after the fact), it may be possible to increase the likelihood of experiencing resilient, if not growth, outcomes in Antarctic expeditioners.

A proposed model integrating these factors as they relate to Antarctic populations is presented in Figure 3.2, with each component being discussed in greater detail below. Although the emphasis and exact nature of experiences within each category within the model dif-

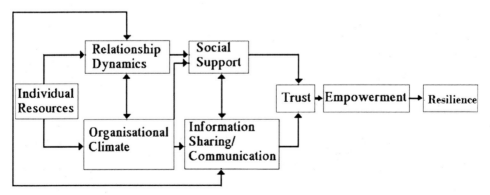

Figure 3.2. Mechanisms underpinning resilient outcomes in Antarctic poplations.

fered between expeditioners and partners, the high degree of similarity across the overarching themes enabled the development of a single model to account for the experiences of expeditioners and partners alike.

Individual Resources

Individual resources referred to preexisting psychological and physiological resources that facilitated engagement in both relationship and organizational domains. In particular, it was identified that optimism, self-efficacy, and the use of positive reframing all consistently enhanced adaptive capacity throughout the employment experience for all participant categories. Research indicates that these individual resources identified within the present thesis (i.e., optimism, self-efficacy, and positive reframing) are not static (e.g., Beck & Strong, 1982; Folkman & Maskowitz, 2000), but amenable to change, and therefore may represent important avenues for proactive intervention strategies.

Relationship Dynamics

The importance of relationship dynamics in facilitating resilient outcomes at all stages of the Antarctic employment experience emphasizes the importance of incorporating a comprehensive assessment approach (i.e., beyond the individual) in this context. Positive relationship dynamics such as open communication, communal activities, and agreement on relationship roles facilitate the development of a shared relationship schema (McCubbin, McCubbin, Thompson, Han, & Allen, 1997), which in turn facilitates shared positive experiences, trust in partners/significant others, and empowerment as a result of these.

The presence of a close, confiding relationship has been associated with lower levels of distress (Dunkel-Schetter & Bennett, 1990; Wethington & Kessler, 1986) and greater resilience following challenging events due to collective cognitive convergence (Parunak, et al., 2008). Fischer and Phillips (1982) further argue that individuals who lack intimate relationships are disadvantaged with regard to access to social support, with this likely to be due to the absence of a readily accessible and reciprocal confiding relationship (Dean, Lin, Tausig, &

Ensel, 1980; Miller & Ingram, 1979). Relationships provide the social context in which collective knowledge can be shared and needs articulated (Paton et al., 2008). For this reason, individuals within these relationships need to be able to accurately articulate their problems and communicate questions about these problems to organizational agencies so they can receive the information that is relevant to their needs in order to have a more accurate perception regarding the nature of current and future challenges (Lasker & Weiss, 2003). This will reduce their uncertainty and increase their trust in the organization.

Organizational Climate

Organizational climate was identified as a predictor of expeditioner resilience at all stages of the Antarctic employment experience. Considering that they are consistently engaged in this culture, with little demarcation between work and non-work roles limited during the absence period, this is not surprising. Partners expressed a desire to have more influence and engagement with the organization when in the presence of the expeditioner (i.e., at predeparture, reunion, and reintegration), which may be precipitated by expeditioner behaviors reminding the partner of challenges associated with the experience and these behaviors being externalized to the organizational climate in which they are working.

Quality supervisor-subordinate relationships, of which supportive supervisor behavior is a crucial factor (Liden, Sparrow, & Wayne, 2000), creates the conditions necessary for the personal growth of individuals (Cogliser & Schriesheim, 2000), enhancing general feelings of competence. Additionally, quality supervisor-subordinate relationships encourage the creation of similar value structures between employees (Cogliser & Schriesheim, 2000), building shared schema, enabling employees to find increased meaning in their task activities, and contributing to the development of a sense of cohesion and trust between colleagues. Within the current study, expeditioners who perceived a positive relationship with the station leader were more likely to report themes consistent with empowerment than those who did not. Additionally, findings from the current research suggest that organizational climate extends beyond the employer-employee relationship to encompass those indirectly affected (i.e., partners). Partners also need to be involved in the organizational environment to facilitate effective communication exchange and the development of a schema which integrates aspects of the organization in a meaningful way in order to experience empowerment and resilience throughout the Antarctic employment experience.

Social Support

Social support was identified as a factor facilitated by positive relationship dynamics, positive organizational climate, and information sharing/communication. Ready access to social support appeared to serve both a supportive and a normalizing function. Within the present research, social support referred to the provision of support (informal and formal) to the participant by external agents. There is an extensive body of evidence that testifies to the importance of quality social support in negotiating challenging experiences. For example, it has been demonstrated that access to social and information resources allows individuals to take initiative and enhance their sense of control (impact) and self-efficacy (competence) over challenges (Gist & Mitchell, 1992; Paton, 1994), thereby contributing to feelings of empowerment and engendering resilient outcomes (Paton et al., 2008). In turn, provision of social support between colleagues enhances team cohesiveness. Members of cohesive work teams are more willing to share their knowledge and skills, an essential prerequisite for the development and maintenance of the learning culture that is fundamental to agency and officer resilience (Paton et al., 2008).

Information Sharing/Communication

Within the current research, information sharing/communication referred to the degree to which information dissemination occurred between the organization and the participants (both expeditioners and partners) as well as the degree of information dissemination occurring between expeditioners and partners. Relationships in which quality communication and information sharing occurs has been demonstrated to facilitate empowerment and thereby adaptation (Walsh, 2006). Sharing of clear information and messages, open empathic emotional sharing, and collaborative problem solving between both intimate and collegial supports engender feelings of individual mastery and self-efficacy over challenging situations, increase trust in others to provide support in response to these situations, and facilitate feelings of empowerment and resultant adaptation (Walsh, 2006).

Information sharing/communication was facilitated by ready access to social support as well as positive relationship dynamics and organizational climate. The process of information sharing/communication facilitated resilience in two ways: (1) greater knowledge and understanding of the situation enabled greater feelings of control and sense of mastery over the situation (factors routinely associated with the development of positive adaptation; Miller & Kaiser, 2001), and (2) facilitating the development of closer relationships with peers, thereby enabling ready access to social support. Previous research has further identified that an important resource that plays a pivotal role in predicting empowerment is information sharing (Paton et al., 2008) through creating a sense of clarity regarding the situation (Paton & Flin, 1999) as well as purpose and meaning (Conger & Konungo, 1988) among all involved parties. However, information is not enough. The social context in which information is received is an equally important determinant of empowerment (Paton et al., 2008). In this context, one aspect of the agency-officer relationship becomes particularly important, and that concerns trust.

Trust

Within the current research findings, trust referred to the degree to which participants believed that they had the ability to successfully negotiate the Antarctic employment experience and experience positive outcomes as a result in both the short and long term. It also referred to the belief that both intimate partners and the organization would support them in this task.

Previous research has determined trust to be an important predictor of the effectiveness of interpersonal relationships (including both intimate and platonic relationships), group processes, and organizational relationships (Barker & Camarata, 1998; Herriot, Hirch, & Reilly, 1998), and plays a crucial role in empowering individuals (Paton et al., 2008; Spreitzer & Mishra, 1999). People functioning in trusting, reciprocal relationships are left feeling empowered and more likely to experience meaning in their work. Trust has been identified as a predictor of people's ability to deal with complex, high risk events

(Paton et al., 2008; Siegrist & Cvetkovich, 2000), particularly when relying on others to provide information or assistance. Trust influences perception of others' motives, their competence, and the credibility of the information they provide (Earle, 2004). An individual is more willing to commit to undertaking a challenging task (such as Antarctic employment) when he or she believes those with whom he or she must collaborate or work under are competent, dependable, likely to act with integrity (in the present and in the future), and care for his or her interests (Dirks, 1999). Organizations functioning with cultures that value openness and trust create opportunities for employees to engage in learning and growth, contributing to the development of adaptive capacity (Barker & Camarata, 1998; Siegrist & Cvetkovich, 2000).

Empowerment

Within the current research, empowerment was evidenced by participant reports of enhanced competence, the development of new skills sets, increased self-efficacy, and feelings of personal involvement and control over meaningful outcomes. Previous researchers have identified that empowerment predicts satisfaction in individuals and teams (Kirkman & Rosen, 1999; Koberg, Boss, Senjem, & Goodman, 1999), which may assist in explaining its relationship to resilience within the current research. However, Paton et al. (2008) argued that empowerment has demonstrated strong links to motivating action in conditions of uncertainty (Conger & Konungo, 1988; Spreitzer, 1997) that renders it capable of providing valuable insights into resilience and adaptation, as well as how these outcomes can be developed and sustained.

Motivational interpretations of empowerment derive from a theoretical perspective that argues that if people have sufficient resources (psychological, social, and physical) as well as the capacity to engage them, they will be able to effectively confront challenges presented by events (Conger & Konungo, 1988; Spreitzer, 1996), such as Antarctic employment. In this way, theories of empowerment readily integrate individual, relationship, and organizational factors such as that demonstrated within Figure 2. Conger and Konungo (1988) conceptualize empowerment as an enabling process that facilitates the conditions necessary to effectively confront (i.e., develop meaning and competence) future challenges. Conger and Konungo argue that individual differences in meaning and competence reflect the degree to which the environment (i.e., the organizational climate) enables actions to occur.

CONCLUSION

The majority of expeditioner and partner responses indicated that there are minimal long-term negative effects associated with Antarctic employment experiences. In fact, there was evidence of improved functioning beyond that reported at predeparture consistent with not only resilience but also growth outcomes. A growing body of evidence exists reporting the coexistence of both positive and negative outcomes and how these relate to positive and negative change (e.g., Tedeschi & Calhoun, 1995). In this way, results of the current study further emphasize the need to adopt comprehensive approaches to research undertaken with Antarctic populations, including adoption of a salutogenic paradigm that acknowledges the possibility (if not likelihood) of concurrent positive and negative experiences and resilience and vulnerability factors interacting to influence outcomes.

Furthermore, it would appear that the high levels of traits associated with resilience and growth outcomes reported at predeparture suggest an ability to adapt to challenging situations largely by facilitating a cognitive processing style that maximizes positive input as opposed to negative input. Although some authors have argued that this propensity is related to innate personality characteristics, it is possible that by addressing factors identified as predictive of these traits (e.g., individual- and relationship-level coping strategies, collaborative organizational practices; Norris, 2010), functioning may be maximized for expeditioners and partners alike. In turn, by maximizing the positive outcomes associated with the Antarctic experience, it may be possible to enhance the retention of experienced expeditioners for future employment.

The model as presented in Figure 3.2 shares a number of commonalities with the Stress-Shield model of resilience as proposed by Paton et al. (2008). The Stress-Shield model of resilience was developed to explain the processes contributing to resilience in police officers and proposes that resilience reflects the extent to which individuals and the groups to which they belong can capitalize on resources and competencies (both psychological and physical) in ways that allow challenging events to be rendered coherent, manageable, and meaningful (Paton et al., 2008).

Similarly, the current model (Figure 3.2) was developed to explain the processes contributing to resilience within Antarctic populations and also reflects the importance of capitalizing on both individual- and group-level competencies in order to process, organize, and comprehend challenging experiences in a way that facilitates resilient outcomes. Specifically, both models emphasize the multidimensional nature of resilience by incorporating individual and organisational factors, as well as the influence of trust, the valence of experience (influenced by cognitive constructions of events), and empowerment suggesting that despite differences in the nature of experiences and contexts in which they occur, there may be overarching universal factors (e.g., empowerment) that contribute to resilient outcomes.

However, the current model (Figure 3.2) emphasizes information sharing/communication and social support to a greater degree than the Stress-Shield model, suggesting differences in underlying processes that facilitate positive outcomes as a function of vocational experiences and providing further evidence for the argument that vocationally specific models of resilience should be developed to ensure provision of appropriate interventions and supports. Additionally, the current model expands on the Stress-Shield model of resilience to incorporate relationship factors and thereby accommodates the demonstrated influence of relationship dynamics on individual well-being.

At this time, the proposed model requires empirical validation through the use of structural equation modelling techniques before guiding intervention practices. Additionally, considering that 20 nations maintain a constant human presence in Antarctica through undertaking scientific research at permanent stations, it is imperative that future research assess the cross-cultural applicability of the proposed model.

REFERENCES

Barker, B. T., & Camarata, M. R. (1998). The role of communication in creating and maintaining a learning organization: Preconditions, indicators, and disciplines. *Journal of Business Communication, 35,* 443–467.

Beck, J. T., & Strong, S. R. (1982). Stimulating therapeutic change with interpretations: A comparison of positive and negative connotation. *Journal of Counselling Psychology, 29,* 551–559.

Cogliser, C. C., & Schriesheim, C. A. (2000). Exploring work unit context and leader-member exchange: A multilevel exchange. *Journal of Organisational Behaviour, 21,* 487–511.

Conger, J. A., & Konungo, R. N. (1988). The empowerment process: Integrating theory and practice. *Academy of Management Review, 13,* 471–482.

Dean, A., Lin, N., Tausig Lin, N., Taursig, A., & Ensel, W. M. (1980). The epidemiological significance of social support systems in depression. In R. G. Simmons (Ed.), *Research in community and mental health* (Vol. 2). Greenwich, CT: JAI Press.

Decamps, G., & Rosnet, E. (2005). A longitudinal assessment of psychological adaptation during a winter-over in Antarctica. *Environment and Behaviour, 37,* 418–435.

Dirks, K. T. (1999). The effects of interpersonal trust on work group performance. *Journal of Applied Psychology, 84,* 445–455.

Dunkel-Schetter, C., & Bennett, T. L. (1990). Differentiating the cognitive and behavioural aspects of social support. In B. R. Sarason, I. G. Irwin, & G. R. Pierce (Eds.), *Social support: An interactive view* (pp. 267–296). Oxford, England: John Wiley and Sons.

Earle, T. C. (2004). Thinking aloud about trust: A protocol analysis of trust in risk management. *Risk Analysis, 24,* 169–183.

Fischer, C. S., & Phillips, S. L. (1982). Who is alone? Social characteristics of people. In L. A. Peplau & D. Perlman (Eds.), *Loneliness* (pp. 21–39). New York: Wiley.

Folkman, S., & Maskowitz, J. T. (2000). Positive affect and the other side of coping. *American Psychologist, 55,* 647–654.

Fredrickson, B. L. (2003). The value of positive emotions. *American Scientist, 91,* 330–335.

Gist, M. E., & Mitchell, T. R. (1992). Self-efficacy: A theoretical analysis of its determinants and malleability. *Academy of Management Review, 17,* 183–211.

Herriot, P., Hirsch, W., & Reilly, P. (1998). *Trust and transition. Managing today's employment relationship.* New York: Wiley.

Houtzager, B. A., Oort, F. J., Hoekstra-Weebers, J. E., Caron, H. N., Grootenhuis, M. A., & Last, B. F. (2004). Coping and family functioning predict longitudinal psychological adaptation of siblings of childhood cancer patients. *Journal of Paediatric Psychology, 29,* 591–605.

Janoff-Bulman, R. (1992). *Shattered assumptions: Towards a new psychology of trauma.* New York: Free Press.

Kirkman, B. L., & Rosen, B. (1999). Beyond self-management: Antecedent and consequences of team empowerment. *The Academy of Management Journal, 42,* 58–74.

Koberg, C. S., Boss, R. W., Senjem, J. C., & Goodman, E. A. (1999). Antecedents and outcomes of empowerment. *Group and Organisation Management, 24,* 71–91.

Lasker, R. D., & Weiss, E. S. (2003). Broadening participation in community problem solving: A multidisciplinary model to support collaborative practice and research. *Journal of Urban Health, 80,* 14–47.

Liden, R. C., Sparrow, R. T., & Wayne, S. J. (2000). An examination of the mediating role of psychological empowerment on the relations between the job, interpersonal relationships, and work outcomes. *Journal of Applied Psychology, 85,* 407–416.

Lugg, D. J. (2005). Behavioural health in Antarctica: Behavioural implications for long-duration space missions. *Aviation, Space, and Environmental Medicine, 76,* 74–77.

McCubbin, H. I., McCubbin, M. A., Thompson, A. I., Han, S. Y., & Allen, C. A. (1997). Families under stress: What makes them resilient. *Journal of Family and Consumer Sciences, 89,* 2–11.

Miller, C. T., & Kaiser, C. R. (2001). A theoretical perspective on coping with stigma. *Journal of Social Issues, 57,* 73–92.

Miller, P., & Ingram, J. G. (1979). Reflections on the life events to illness link with some preliminary findings. In I. G. Sarason & C. D. Spielberger (Eds.), *Stress and anxiety* (Vol. 6, pp. 313–336). New York: Hemisphere.

Norris, K. (2010). *Breaking the ice: Developing a model of expeditioner and partner adaptation to Antarctic employment.* Unpublished doctoral dissertation, University of Tasmania, Australia.

Palinkas, L. A. (2003). The psychology of isolated and confined environments: Understanding human behaviour in Antarctica. *American Psychologist, 58,* 353–363.

Palinkas, L. A., & Suedfeld, P. (2007). Psychological effects of polar expeditions. *Lancet, 371,* 153–163.

Parunak, H. V., Belding, T. C., Hilscher, R., & Brueckner, S. (2008, May 12–16). Modeling and managing collective cognitive convergence. *Proceedings of 7th International Conference on Autonomous Agents and Multiagent Systems (AAMS),* Estoril, Portugal.

Paton, D. (1994). Disaster relief work: An assessment of training effectiveness. *Journal of Traumatic Stress, 7,* 275–288.

Paton, D. (2006). Posstraumatic growth in disaster and emergency work. In L. G. Calhoun & R. G. Tedeschi (Eds.), *Handbook of posttraumatic growth: Research and practice* (pp. 225–247). Mahwah, NJ: Lawrence Erlbaum Associates.

Paton, D., & Flin, R. (1999). Disaster stress: An emergency management perspective. *Disaster Prevention and Management, 8,* 261–267.

Paton, D., Violanti, J. M., Johnston, P., Burke, K. J., Clarke, J., & Keenan, D. (2008). Stress shield: A model of police resiliency. *International Journal of Emergency Mental Health, 10,* 95–108.

Patterson, J. M. (2002). Integrating family resilience and family stress theory. *Journal of Marriage and the Family, 64,* 349–360.

Siegrist, M., & Cvetkovich, G. (2000). Perception of hazards: The role of social trust and knowledge. *Risk Analysis, 20,* 713–720.

Southwick, S. M., Vythilingam, M., & Charney, D. S. (2005). The psychobiology of depression and resilience to stress: Implications for prevention and treatment. *Annual Review of Clinical Psychology, 1,* 255–291.

Spreitzer, G. M. (1996). Social structural characteristics of psychological empower-ment. *Academy of Management Journal, 39,* 483–504.

Spreitzer, G. M., & Mishra, A. K. (1999). Giving up control without losing control: Trust and its substitutes' effects on managers involving employees in decision making. *Group Organisation Management, 24,* 155–187.

Stinnet, N., & DeFrain, J. (1985). *Secrets of strong families.* Boston: Little, Brown.

Suedfeld, P. (1991). Polar psychology: An overview. *Environment and Behaviour, 23,* 653–665.

Taylor, A. J. W. (1973). The adaptation of New Zealand personnel in the Antarctic. In O. G. Edholm & E. K. E. Gunderson (Eds.), *Polar human biology: The proceed-ings of the SCAR/IUPS/IUBS Symposium on Human Biology and Medicine in the Antarctic* (pp. 417–429). Great Britain: William Heinemann Medical Books Ltd.

Tedeschi, R. G., & Calhoun, L. G. (1995). *Trauma and transformation: Growing in the aftermath of suffering.* Newbury Park, CA: Sage.

Walsh, F. (1996). The concept of family resilience: Crisis and challenge. *Family Process, 35,* 261–281.

Walsh, F. (2003). Family resilience: Strengths forged through adversity. In R. Walsh (Ed.), *Normal family processes* (3rd ed., pp. 399–423). New York: Guilford Press.

Walsh, F. (2006). *Strengthening family resilience* (2nd ed). New York: Guilford Press.

Wethington, E., & Kessler, R. C. (1986). Perceived support, received support, and adjustment to stressful life events. *Journal of Health and Social Behaviour, 27,* 78–89.

Wood, J., Hysong, S. J., & Lugg, D. J. (2000). Is it really so bad? A comparison of positive and negative experiences in Antarctic winters stations. *Environment and Behaviour, 32,* 84–110.

Chapter 4

BUSINESS RESILIENCE IN THE FACE OF CRISIS AND DISASTER

Douglas Paton and John McClure

INTRODUCTION

When disaster strikes, the capacity of businesses to continue their activities and recover from the losses and disruption experienced is important not only to the existence of the business itself but also for the socioeconomic vitality of areas affected by a disaster. Natural disasters can have a huge impact on businesses. For example, following the Loma Prieta (San Francisco) earthquake in 1989, the businesses that suffered the most disruption in the Northridge, California, earthquake were least likely to recover in the subsequent three years (Dahlhamer & Tierney, 1996). Levene (2004) discussed how a lack of business preparedness accounted for about 25 percent of the $40 billion lost as a result of the September 11 terrorist attacks in New York. Levene also pointed out that an estimated 90 percent of medium to large companies that can't resume near-normal operations within five days of an emergency face substantially increased risk of going out of business within five years. These problems may be considerably more acute for small businesses. Businesses must plan to consider how they will contend with, for example, the temporary or permanent loss of employees (e.g., unable to get to work, injured, employees psychologically incapacitated by stress and traumatic stress), production and distribution problems, temporary or permanent inability to access premises, loss of utilities, and changes in patterns of consumption.

A key element of business resilience is recovery planning. This facilitates a capacity for bouncing back by guiding focusing attention on planning how a business may return to predisaster levels of functioning. In doing so, it is important that business planning considers how it can assist its employees. Businesses are social settings. As such they can provide structured and informal social support for employees (Paton, 1997a). None of these activities and benefits can arise unless the business is prepared and has anticipated how it will respond to a crisis. Anticipating what could happen and planning to respond is essential (Coutu, 2002).

Business resilience is built on planning: disaster management plans, recovery plans, and communication plans. It includes clearly identifying the roles of key personnel, rehearsing plan activation, training and updating plans in response to feedback, and having recovery plans that encompass global employee assistance programs, on-site counseling, peer support programs, and telephone crisis counselling (Paton, 1997b). Planning commences with business impact analysis and risk identification in the event of a disaster.

This includes breaking down issues to identify how a business may be affected and using these data to guide the development of response plans and strategies to facilitate continued operations during and after disasters (Elliott, Swartz, & Herbane, 2002; Rose & Lim, 2002). Disaster recovery plans minimize disruption to key elements identified in the business impact analysis and risk assessment and provide the foundation for business continuity management.

Business continuity management (BCM) describes how businesses can assess risk and develop plans and strategies to mitigate the risks posed by hazardous events (Shaw & Harrald, 2004). This chapter discusses how this process can be used to develop the capacity of businesses to sustain organizational activity in the event of disaster.

BUILDING RESILIENCE INTO BUSINESS ACTIVITY

Dahlhamer and D'Souza (1995) examined several predictors of business preparedness in the United States: business size, business age, type of business, whether it was individual or a franchise, and whether it was owned or leased. They found that the business characteristics model (size and type of business) was a significant predictor of preparedness. The size of the business, defined in terms of the number of full-time employees, was the strongest predictor. Larger companies are more prepared (see also Webb, Tierney, & Dahlhamer, 2000), and in many cases larger companies have a business continuity plan (see below), whereas smaller companies do not. In one of the two counties, the type of business was a predictor, in that finance, insurance, and real estate sectors were more prepared than other sectors. This may reflect the likelihood that businesses in these industry sectors have financial risk management procedures in place, making it easier for them to adapt plans to accommodate risk from other sources.

Ownership was a significant predictor. Businesses that were part of a national chain of businesses were more prepared than firms run by sole proprietors. Interestingly, the company's financial situation was unrelated to preparedness as an individual predictor, although this factor may have acted indirectly through business size. The authors concluded that incentives and regulation are necessary to achieve high levels of preparedness as the level of voluntary preparedness in businesses was so low. Businesses with prior experience of a natural disaster tend to be more prepared (Webb et al., 2000). In some regions, preparedness also related to the type of business; for example, in Northridge and Memphis, the finance, insurance, and real-estate sectors were more prepared than retail and service areas. This finding highlights the importance of tailoring outreach activities to the needs of different sectors in order to target their specific needs, rather than sending the same undifferentiated message to all types of companies.

Yoshida and Delye (2005) examined predictors of Florida businesses carrying out mitigation actions, having a business continuity plan (BCP), and purchasing insurance. A regression analysis found three significant predictors of businesses' decision to employ mitigation measures. The strongest predictor was access to expertise such as a structural engineer; the second predictor was the type of business, in that businesses in education, social services, finance, insurance, and real estate had higher levels of preparedness and a BCP than other businesses. In some cases, this difference may be because many of the companies in these sectors are required by their national or state professional organizations to have a business continuity plan. In contrast, businesses in engineering, architecture, and accounting had carried out more mitigation measures. The third significant predictor was their perceived exposure to natural hazards—a measure of companies' perception of the risk. As in other studies, small companies were less prepared as were businesses that were run from the owner's home. Consequently, business education should target small businesses, those run from homes, and those types of business where preparation is lower.

PREVENTING LOSS VERSUS FACILITATING SURVIVAL

When it comes to what to do, business are more likely to undertake survival actions (e.g., purchasing a medical kit) than take actions to reduce loss of buildings (McClure et al., 2007). A failure to take mitigation actions increases the risk of the loss of premises and a need to find alternatives in the recovery phase. Yoshida and Deyle (2005) examined what factors influence small businesses adopting hazard mitigation measures. They found that access to expertise (e.g., insurance manager, structural engineer, businesses continuity specialist, and disaster recovery specialist) was a significant predictor of businesses' decision to employ measures to mitigate damage, as were perceived exposure to natural hazards and type of business. Businesses in education, social services, finance, insurance, and real estate were more likely to have a BCP in place, and perceived exposure to hazards was positively related to structural mitigation and insurance purchase.

One reason for businesses preferring survival actions over mitigation measures could be the difference in their relative costs (Webb et al., 2000). Yoshida and Delye (2005) similarly found that the three most commonly adopted measures in a sample of businesses in Florida cost less than $500. However, this difference in performance of the actions is maintained even with regard to the performance of mitigation actions that are low in cost, such as fitting computer restraints or securing bookshelves (McClure et al., 2007). For example, in the 1989 Loma Prieta (San Francisco) earthquake, the strongest predictor of business resilience was whether businesses had undertaken mitigation actions and in particular whether companies had used computer restraints or backups, which involve a relatively low cost (Yoshida & Delye, 2005).

The prior discussion of how cost issues may influence decision making introduces a need to consider whether things such as incentives, rewards, or regulations are required to motivate actions. McClure,

Fischer, Charleson, and Spittal (2009) concluded that to get companies to spend time and money to take action, information needs to be reinforced by incentives and/or regulation.

Motivation is often construed in terms of the costs and benefits relating to a given action. A cost-benefit analysis is often employed in businesses to assess whether the expected benefits from a proposed action exceed its expected costs. If the total value of the costs exceeds the total value of the benefits, the relevant action should not be taken.

In targeting hazard preparedness in terms of costs and benefits, it is important to counter the perception that only major expenditures are useful in mitigating damage from huge events like floods or earthquakes. In the Loma Prieta earthquake, the strongest predictor of business survival was whether companies had a computer lock. Businesses that had no computer locks lost all their data and records and took months to regain some or all of these data, often going bankrupt as a consequence. In contrast, businesses that had secure computers were able to function the following day or week, even if in some cases they had to shift premises due to damage to their building. In this case, the very small cost of the computer security had huge benefits. With large companies, this security involves servers and other backups rather than computer locks, but the principle still holds because these security measures are relatively inexpensive. Thus, it is important to communicate that large benefits do not necessarily entail large costs.

The adoption of survival and mitigation actions is one aspect of business resilience. Another concerns the actions required to facilitate the ability of the business to function over time in the context of the kinds of losses and disruption that businesses could encounter in the aftermath of a disaster. This is the function of business continuity planning.

BUILDING RESILIENCE:
THE ROLE OF CONTINUITY PLANNING

Business continuity planning and management encompasses practices and strategies that facilitate the ability of a business to cope with, adapt to, and recover from crises such as confronting the consequences of natural disaster (Comfort, 1994; Paton, 2006; Shaw & Harald, 2004). It provides an overarching framework for the systematic appraisal of the conditions that could confront the business and guides the development of the systems and competencies required to facilitate the continuity of business activity under atypical crisis conditions. It can thus confer on a business and its employees a capacity to maintain levels of functioning during and following a disaster, adapt to changes in their circumstances that arise from hazard consequences, and hasten the return of the business to normal functioning during the recovery period.

Pursuing this objective requires businesses to attend to several issues. First, it is important to implement and develop appropriate survival and mitigation actions (see above). Second, it requires that management and information systems are available (by safeguarding existing systems and/or arranging for substitutes) to facilitate continuity of core business operations (Davies & Walters, 1998; Duitch & Oppelt, 1997; Lister, 1996). Third, it requires the development, testing, and regular updating of management systems and procedures designed specifically for managing crisis and for managing the transition between routine and crisis operations in the context of the kind of losses and disruption that the business may have to contend with in the event of their experiencing a large-scale natural disaster (Paton, 1997a, 1997b; Shaw & Harrald, 2004).

Disaster associated with natural hazard activity (e.g., large-scale earthquakes) represents the upper end of the scale of events that need to be considered within the business continuity planning process (Reiss, 2004). Although businesses often prepare for small-scale disruptions (e.g., loss of power supply for a few days), planning for such low-level eventualities will not confer on a business a capacity to respond to more significant events (Paton, 1997a, 1997b). Qualitative differences in the nature, scale, and duration of impacts mean that planning for routine losses represents an inappropriate basis for disaster business recovery planning. By planning for large-scale disasters, BCPs will be able to accommodate the impact of lesser events (e.g., loss of utilities). When planning for large-scale disasters, businesses must consider how they will deal with, for example, prolonged loss of utilities (e.g., power, water, gas), conducting core operations away from their headquarters, dealing with casualties and deaths among staff, reconciling work with the family needs and concerns of staff, ensuring that appropriate crisis management systems and procedures are in place, and ensuring that staff fulfilling disaster continuity roles can deal with high demands over prolonged periods of time (possibly over several months).

The last point illustrates how continuity planning involves ensuring the availability of staff capable of operating crisis management systems under challenging circumstances (Shaw & Harrald, 2004). Staff and managers responsible for implementing BCPs must thus be specifically selected and trained for these roles. This is an essential prerequisite to the development of plans and their effective implementation in the event of a crisis.

Developing Continuity Plans and Strategies

BCM is a proactive and holistic management process designed to provide an iterative, structured process that incorporates risk identification, assessment and management, the development of disaster recovery plans and procedures, training, exercising, and using feedback from exercises to promote the iterative development of plans and capabilities. BCM is built around understanding what the organization must achieve (its critical objectives), identifying the barriers or interruptions that may prevent their achievement if a business experiences large-scale natural hazard consequences, and planning how the business and its employees will act to ensure that core business objectives can be pursued should disruption occur (Elliott et al., 2002; Business Continuity Institute, 2002).

Staff participation is fundamental to effective BCM. It helps embed continuity planning into the culture of the organization, contributes to staff morale by heightening awareness that the business is concerned with their welfare, facilitates communication between key groups and individuals within the business, and provides procedures for risk assessment and mitigation. To accommodate changes in personnel, business practices, or demands from the external environment (e.g., supplies, infrastructure and transportation losses, etc.) over time, the plan should be tested, maintained, and revised regularly.

Implementation

All businesses are different. Consequently, BCM plans and activities must be tailored to the needs of each business. An important issue in this context is ensuring that the BCM includes a check list of "who does what" in the event of a disaster and identifies practical strategies for cooperating with, for example, the emergency services, suppliers and customers, the utility companies, local authorities, the insurance companies, and perhaps other businesses in the area.

Implementation planning should also include the adequate allocation of resources, both financial and human. Consequently, managerial acceptance of risk and their commitment to BCM is essential to planning being initiated and developed to an appropriate state of readiness. The culture of the organization, particularly with regard to its approach to strategic change, plays a role in determining whether this need is identified and acted on. The development of an appropriate change and risk management culture is a challenging task, especially for small businesses. First, it is pertinent to consider how the beliefs of managers and other key players can influence this process.

Organizational commitment to disaster BCP can be constrained by managers overestimating existing capabilities and by ambiguity of responsibility (Folke, Colding, & Berkes, 2003; Gunderson, Holling, & Light, 1995; Shaw & Harrald, 2004). Issues regarding responsibility are particularly important. In large organizations, because continuity planning crosses several organizational role boundaries, responsibility for its performance may not fall within the purview of any one established organizational role. In small businesses, preoccupation with day-to-day activities can preclude considering continuity planning. Consequently, a precursor to effective BCM is having responsibility vested in a key figure who can direct and sustain the planning process (Shaw & Harrald, 2004). For small business, entities such as Chambers of Commerce and industry groups could help facilitate this process. Preparing plans and developing organizational capability is one important part of the process. The other is ensuring the availability of staff capable of implementing plans under atypical crisis conditions.

BUILDING CAPABILITY IN ORGANIZATIONS

Building organizational capability requires a substantial commitment from management, particularly with regard to pledging the financial resources, time, and effort required to prepare for something that could occur tomorrow but may not occur for several years, if ever.

The costs of BCM are highly visible, the benefits less so. This may be less of an issue for businesses that already have effective risk management policies because these can provide a springboard for BCM. However, for other organizations, particularly small businesses, ensuring owner or management commitment is more difficult. Increasingly, though, pressure for this to happen is coming from stakeholders (e.g., suppliers, customers, employees) who want to know that the organization is prepared to deal with crises so that their investments and livelihoods are protected. Larger organizations that have implemented BCM are putting pressure on suppliers to protect themselves from any breakdown in the supply chain. Smaller businesses may, however, need assistance (financial and expertise) to put plans and competencies in place. Making the decision to implement BCM is one thing; organizations then need to implement the necessary changes to culture, attitudes, and practices.

Learning and Change in Contemporary Organizations

Developing disaster BCP requires thinking about nonroutine events and possibilities. Planning in this context of change requires understanding how managers think about future, nonroutine events about which there is considerable uncertainty. This task is complicated by the fact that, over time, the conceptual frameworks or "mental maps" that inform managers' thinking and action become entrenched in routine activities and can become insulated from the environmental inputs. This can result in managers becoming "cognitively complacent." This can result in managers assimilating the unknown by attempting to make sense of new, complex, and ambiguous environmental data (such as that required to understand potential impacts of hazards on business activity) by making them "fit in" with previous experience rather than anticipating the need for alternative ways of thinking and acting (Paton & Wilson, 2001).

This tendency makes it difficult for managers to consider, far less confront, the nonroutine BCM contingencies they may have to contend with during and after a disaster. Consequently, managers responsible for BCM planning must engage in a level of environmental monitoring, discussion with others (e.g., scientific and emergency management agencies), and creative decision making that is unique to this activity. By understanding the cognitive processes that guide strategic thinking, managers can engage in activities that challenge assumptions, facilitate change, develop the ability to anticipate the unforseen, and take steps to mitigate the action of factors that adversely affect organizational willingness and/or ability to change.

Several factors can lead organizations to deny their vulnerability to loss and disruption from potential disasters. These include thinking that crises only happen to other organizations or that the organization is too big and powerful to be affected by a disaster (Mitroff & Anagnos, 2001). Such beliefs can prevent change or render the planning and implementation of change into a more challenging endeavour. Implementing change can be particularly problematic for organizations where power and authority are highly centralized (Gunderson et al., 1995; Harrison & Shirom, 1999). In contrast, businesses that have sufficient structural flexibility are in a better position to develop their capability to manage significant disruptions (Alesch, Holly, Mittler & Nagy, 2002; Folke et al., 2003; Paton, 1997a, 1997b).

Folke et al. (2003) emphasized the fact that, to increase resilience, businesses can benefit from having had some experience of failure. But to benefit, they need to have an ability to learn from it. Failure provides valuable insights into areas where development is required. The idea that a business should plan for failure as well as success is a difficult concept to accept. However, "failing to plan to fail" is as important as "failing to plan to succeed" (Folke et al., 2003). Finally, these lessons must be encapsulated in new policies and procedures that facilitate organizational adaptability (Coutu, 2002; Folke et al., 2003; Paton & Wilson, 2001). One important outcome of this process is identification of the competencies and capabilities required of the staff who will be responsible for implementing the plan during a disaster.

DEVELOPING EFFECTIVE CAPABILITY IN STAFF

The environment within which businesses will be required to operate during and after a disaster will differ substantially from that in which routine business activity is undertaken. If the benefits of BCM planning are to be fully realized and can be implemented in a timely and effective manner, staff capable of applying them in a context defined by a need to confront challenging circumstances must be developed. This can be accomplished by including selection and training activities in the planning process.

Selection and Training

In addition to selecting for specific competencies (e.g., operational skills, crisis decision making), it is important, if possible, to select for a capability to deal with the high stress environment in which continuity plans will unfold as the business faces up to the challenges posed by the crisis event (Paton, 1997a; Scotti et al., 1995). Staff can be selected for dispositional resilience factors such as, for example, hardiness, emotional stability, decisiveness, controlled risk taking, self-awareness, tolerance for ambiguity, and self-efficacy (Flin, 1996; Lyons, 1991; MacLeod & Paton, 1999; Paton, 1989, Paton, 2003; Paton & Jackson, 2002).

Organizations may not, however, have the luxury of selecting staff in this manner. There may be insufficient flexibility to afford an opportunity to implement this option or staff may be cast into crisis roles by the unexpected timing of the crisis event. Under these circumstances, knowledge of predictors of stress vulnerability and resilience can be used for the postevent assessment of staff to identify those at risk and to prioritize them for support and monitoring during and after the disaster (Lyons, 1991; Paton, 1989; Tehrani, 1995). Once selected for these roles, staff need to be trained.

The effectiveness of plan implementation will be a function of staff having the competencies (e.g., information management) and capabilities (e.g., stress resilience) required to respond in atypical and challenging circumstances (Grant, 1996; Paton, 1997a). To do so, business preparation planning should include training needs analysis conducted explicitly to identify the consequences likely to be encountered during and after a disaster and identifying the competencies required to manage them. Significant differences between routine and postdisaster environments create novel and highly challenging demands for managers (Paton & Jackson, 2002). Training should cover, for example, risk assessment, developing crisis management skills, developing information and decision management skills, and developing crisis management procedures and learning how to implement them in a crisis (operating under devolved authority and planning for transition into crisis roles and from crisis back into routine operations). It is also essential that plans and people's competencies are tested and evaluated using exercises and simulations designed to put training into practice. Simulations afford opportunities for staff to develop the technical and managerial skills required to respond to hazard consequences, practice their use under adverse circumstances, receive feedback on their performance, increase awareness of stress reactions, and rehearse strategies to minimize negative reactions (Flin, 1996; Paton & Jackson, 2002; Rosenthal & Sheiniuk, 1993).

Given that disasters have community-wide consequences, all staff will be affected to some extent. Consequently, training programs for managers should include strategies for facilitating staff recovery and their return to normal functioning and productivity and managing mental health problems. Fulfilling the former involves their acting as good role models (e.g., acknowledging how they have been affected) and providing feedback and information to staff (Paton, 1997a). This behavior demonstrates how to reconcile the personal impact of the event with continuing to work through a crisis or with returning to work after the disaster.

Because it helps staff put their experience into perspective, allows access to support networks, and facilitates employees regaining a sense of control, returning to work is therapeutic and should be encouraged. However, managing the gradual return and reintegration into work requires careful planning and judgment. During the immediate period of people's return to work, managers should ensure that staff do not take on too much too soon and, because cognitive capacities may be temporarily diminished (e.g., as a result of the acute and/or prolonged stress or trauma staff have experienced), remind them to take care when, for example, operating machinery, driving, or making complex decisions. Managers are also well placed to help staff resolve their experiences in a beneficial manner. This can be facilitated by, for example, helping staff to identify strengths that helped them deal with this event and using the experience to focus on developing future capabilities.

Developing resilient staff is one part of this process. Fully realizing the benefits of BCP is a function of the degree to which it is integrated into the culture of the organization. Recognition of the importance of organizational culture emphasizes the fact that developing resilient people does not guarantee the resilience of the organization as a whole (Coutu, 2002). Organizational resilience depends on the culture, structure, and business practices of the organization as a whole. BCM provides a framework for building this resilience into an organization. The development of the kind of adaptive, learning organizational culture introduced above can provide an appropriate context for empowering staff to develop and sustain their ability to cope with impacts, adapt to emergent circumstances, and recover as quickly as possible.

BCP should be developed in a consultative manner to ensure they are familiar to, and accepted by, those staff required to act on them, and they should be driven by the goal of developing the capability to respond effectively to a range of hazard events (Lister, 1996; Paton, 1997a, 1997b; Shaw & Harrald, 2004). Plans should be linked to training programs, resource allocation, and disaster simulation exercises. If not, plan effectiveness will be diminished when put into practice (Paton, 1997a, 1997b). These collaborative activities provide staff with tangible evidence of organizational concern for their welfare, a shared responsibility for recovery (Powell, 1991), help sustain staff loyalty (Bent, 1995), and ensure that planning and action occur within a supportive culture (Coutu, 2002; Paton, 1997b). Organizational culture has another contribution to make. It provides the impetus to recognize a need for crisis management systems and procedures.

ORGANIZATIONAL IMPLICATIONS

Key predictors of the development of a practical capability to respond to disaster are organizational characteristics (e.g., management style and attitudes, reporting, and decision procedures) and bureaucratic flexibility (Doepal, 1991; Folke et al., 2003; Paton, 1997b; Powell, 1991; Turner, 1994). Rigid bureaucracies can, by persistent use of established procedures (even when responding to different and more urgent crisis demands), internal conflicts regarding responsibility, and a desire to protect the organization from criticism or blame, complicating the response process. Effective response involves relaxing normal administrative procedures and replacing them with procedures designed specifically to manage response and recovery (for both staff and productivity) and, most important, accepting organizational ownership of the crisis and its implications (Elliot et al., 2002). That is, the onus is on the business to respond and recover as far as possible using its own resources and capabilities.

Crisis management systems are required to cover, for example, delegation of authority; allocation of crisis response tasks, roles, and responsibilities and the development of appropriate management procedures; and identifying and allocating resources necessary to deal with the crisis, information management, communication and decision management, and liaison mechanisms. Flexibility in these systems is important. They will be required to deal not only with the uncharacteristic demands of the crisis but also atypical demands emanating from dealing with unexpected emergent tasks, unfamiliar people and roles, and frequent staff reassignment (Folke et al., 2003; Paton, 1997a, 1997b). Communication systems, designed to meet the needs of diverse stakeholder and response groups, are required for information access and analysis, defining priority problems, guiding emergency resource needs and allocation, coordinating activities, providing information to managers, staff, and the media, and monitoring staff and business needs (Bent, 1995; Doepal, 1991; Paton, 1997a, 1997b). Moreover, these activities may be required over a period of several months.

The implementation of the steps outlined above culminates in the existence of a BCM framework within which the systems, procedures, and competencies required to facilitate a capacity for business to adapt to disaster are developed and ready to be enacted. This process enhances organizational resilience, reduces its susceptibility to disruption, protects employee well-being, and hastens the return to productive capacity by putting in place integrated systems capable of adapting, over the course of the response and recovery period, to accommodate changing staff and business needs.

REFERENCES

Alesch, D. J., Holly, J. N., Mittler, E., & Nagy, R. (2002). *Organizations at risk: What happens when small businesses and not-for-profits encounter natural disasters.* Retrieve February 14, 2003, from http://www.riskinstitute.org/ptr_item.asp?cat_id=1& item_id=1028

Bent, D. (1995). Minimising business interruption: The case for business continuance planning. In A. G. Hull & R. Coory (Eds.), *Proceedings of the Natural Hazards Man-*

agement Workshop 1995. Institute of Geological and Nuclear Sciences, Lower Hutt, New Zealand.

Business Continuity Institute. (2002). Good practice in business continuity management. *Continuity, 6,* 2.

Comfort, L. K. (1994). Risk and resilience: Inter-organizational learning following the Northridge earthquake of 17 January 1994. *Journal of Contingencies and Crisis Management, 2*(3), 157–170.

Coutu, D. L. (2002). How resilience works. *Harvard Business Review,* pp 46–55.

Dahlhamer, J. M., & D'Souza, M. J. (1995). *Determinants of business disaster preparedness in two U.S. metropolitan areas.* Preliminary Paper No. 24. Disaster Research Center, University of Delaware.

Dahlhamer, J. M., & Tierney, K. J. (1996). *Rebounding from disruptive events: Business recovery following the Northridge Earthquake.* Preliminary Paper No. 242. Disaster Research Centers: University of Delaware.

Davies, H., & Walters, M. (1998). Do all crises have to become disasters? Risk and risk mitigation. *Disaster Prevention and Management, 7,* 396–400.

Doepal, D. (1991). Crisis management: The psychological dimension. *Industrial Crisis Quarterly, 5,* 177–188.

Duitch, D., & Oppelt, T. (1997). Disaster and contingency planning: A practical approach. *Law Practice Management, 23,* 36–39.

Elliott, D., Swartz, E., & Herbane, B. (2002). *Business Continuity Management.* New York: Routledge.

Flin, R. (1996). *Sitting in the hot seat: Leaders and teams for critical incident management.* Chichester: John Wiley & Sons Ltd.

Folke, C., Colding, J., & Berkes, F. (2003). Synthesis: Building resileince and adaptive capacity in social-ecological systems. In F. Berkes., J. Colding, & C. Folke (Eds.), *Navigating social-ecological systems: Building resilience for complexity and change.* Cambridge: Cambridge University Press.

Grant, N. K. (1996). Emergency management training and education for public administration. In R. T. Styles & W. L. Waugh (Eds.), *Disaster management in the US and Canada: The politics, policymaking, administration and analysis of emergency management* (2nd ed). Springfield, IL: Charles C Thomas.

Gunderson, L. H., Holling, C. S., & Light, S. S. (1995). *Barriers and bridges to the renewal of ecosystems and organizations.* New York: Columbia University Press.

Harrison, M. I., & Shirom, A. (1999). Organizational diagnosis and assessment. Newbury Park, CA: Sage Publications.

Klein, G. (1997). Recognition-primed decision making. In C. Zsambok & G. Klein (Eds.), *Naturalistic decision making.* Mahwah, NJ: Lawrence Erlbaum Assciates.

Levene, Lord. (2004). Taming the beast–managing business risk. London: Lloyd's of London.

Lister, K. (1996). Disaster continuity planning. *Chartered Accountants Journal of New Zealand, 75,* 72–73.

Lyons, J. A. (1991). Strategies for assessing the potential for positive adjustment following trauma. *Journal of Traumatic Stress, 4,* 93–111.

MacLeod, M. D., & Paton, D. (1999). Police officers and violent crime: Social psychological perspectives on impact and recovery. In J. M. Violanti & D. Paton (Eds.), *Police trauma: Psychological aftermath of civilian combat.* Springfield, IL: Charles C Thomas.

McClure, J., Fischer, R., Charleson, A., & Spittal, M. J. (2009). *Clarifying why people take fewer damage mitigation actions than survival actions: How important is cost?* Wellington, NZ: Earthquake Commission.

McClure, J., Sutton, R. M., & Wilson, M. (2007). How information about building design influences causal attributions for earthquake damage. *Asian Journal of Social Psychology, 10,* 233–242.

Mitroff, I. I., & Anagnos, G. (2001). *Managing crises before they happen.* New York: AMACOM.

Paton, D. (1989). Disasters and helpers: Psychological dynamics and implications for counselling. *Counselling Psychology Quarterly, 2,* 303–321.

Paton, D. (1997a). *Dealing with traumatic incidents in the workplace* (3rd ed). Queensland: Gull Publishing.

Paton, D. (1997b). Managing work-related psychological trauma: An organisational psychology of response and recovery. *Australian Psychologist, 32,* 46–55.

Paton, D. (2003). Stress in disaster response: A risk management approach. *Disaster Prevention and Management, 12,* 203–209.

Paton, D. (2006). Disaster resilience: Building capacity to co-exist with natural hazards and their consequences. In D. Paton & D. Johnston (Eds.), *Disaster resilience: An integrated approach.* Springfield, IL, Charles C Thomas.

Paton, D., & Jackson, D. (2002). Developing disaster management capability: An assessment centre approach. *Disaster Prevention and Management,* 11, 115–122.

Paton, D., & Wilson, F. (2001). Managerial perceptions of competition in knitwear producers. *Journal of Managerial Psychology, 16,* 289–300.

Powell, T. C. (1991). Shaken, but alive: Organisational behaviour in the wake of catastrophic events. *Industrial Crisis Quarterly, 5,* 271–291.

Reiss, C. L. (2004). *Risk management for small business.* Fairfax, VA: Public Entity Risk Institute.

Rose, A., & Lim, D. (2002). Business interruption losses from natural hazards: Conceptual and methodological issues in the case of the Northridge earthquake. *Environmental Hazards, 4,* 1–14.

Rosenthal, P. H., & Sheiniuk, G. (1993). Business resumption planning: Exercising the disaster management team. *Journal of Systems Management, 44,* 12–16.

Scotti, J. R., Beach, B. K., Northrop, L. M. E., Rode, C. A., & Forsyth, J. P. (1995). The psychological impact of accidental injury. In J. R. Freedy & S. E. Hobfoll (Eds.), *Traumatic stress: From theory to practice.* New York: Plenum Press.

Shaw, G. L., & Harrald, J. R. (2004). Identification of the core competencies required of executive level business crisis and continuity managers. *Journal of Homeland Security and Emergency Management, 1* (Article 1). Available at http://www.bepress .com/jhsem

Standards Australia. (2003). *Business Continuity Management,* HB 221.

Tehrani, N. (1995). An integrated response to trauma in three Post Office businesses. *Work & Stress, 9,* 380–393.

Turner, B. (1994). Causes of disaster: Sloppy management. *British Journal of Management, 5,* 215–219.

Webb, G. R., Tierney, K. J., & Dahlhamer, J. M. (2000). Businesses and disasters: Empirical patterns and unanswered questions. *Natural Hazards Review,* 83–90.

Yoshida, K., & Deyle, R. E. (2005). Determinants of small business hazard mitigation. *Natural Hazards Review, 6,* 1–12.

Chapter 5

SCIENCE ADVICE FOR CRITICAL DECISION MAKING

EMMA E. DOYLE AND DAVID M. JOHNSTON

INTRODUCTION

During and after a large-scale disaster, numerous agencies and advisory bodies can be involved in the response and recovery, guiding the critical decisions of emergency managers and other protective service agencies. This is particularly true for extreme natural hazard events, which pose a threat to life, infrastructure, and business, including volcanic eruptions, tsunamis, hurricanes, floods, and severe weather events. For uncertain and unfamiliar events such as these, science agencies, individuals, and collaborative science advisory groups are commonly called on to provide technical advice to emergency managers about the impacts and outcomes. This advice plays an important role in the planning, mitigation, and response, and thus it is vital that it is effectively communicated to aid the decision-making processes.

In this chapter, we briefly review the emergency management structure of New Zealand and how science advice is incorporated into that process, as well as decision-making processes and methods for coping with uncertainty. We then discuss NZ's 2008 national Civil Defense Exercise Ruaumoko, which highlights the importance of training to build a common understanding across the emergency management and science advisory sectors. Finally, we discuss methods that help build resilience via the formation of scientific advisory groups, train-

ing programs, and by increasing knowledge of potential future eruptions and impacts to reduce uncertainty and enhance decision making.

NATURAL HAZARDS AND EMERGENCY MANAGEMENT

NZ's Ministry of Civil Defense and Emergency Management
(MCDEM) promotes and manages policies and programs for civil
defense and emergency management (MCDEM, 2008b) by concentrating on the four Rs: Reduction, Readiness, Response, and Recovery. Civil defense emergency management planning is a require-
ment of agencies across the nation, as part of the Civil Defense Emergency Management Act of 2002, which has been utilized in recent natural hazard crises such as the 2004 Manawatu Floods and the Magnitude 7.1 Darfield earthquake, Canterbury, on September 4, 2010
(Wood, Robins, & Hare, 2010). One of the criterions for effective disaster management, defined by Quarantelli (1997), is to "have a well-
functioning Emergency Operations Centre (EOC)" (p. 51). NZ's national response to any civil defense emergency or crisis event is led
through a central Emergency Operation Centre (EOC) via MCDEM's
National Crisis Management Center (NCMC), which liaises with and
supports the 16 regional council CDEM groups across NZ. Each of
these operates a Group EOC (GEOC), which coordinates and supports the local council and territorial authority CDEM EOCs located
in cities, towns, and districts (Lee, 2010). These EOCs are central command and control facilities activated during an event to handle the
response of multiple agencies (fire, police, protective agencies, Civil
Defense, volunteers, etc.) via a number of key coordination points following the structure of the Coordinated Incident Management System
(NZ Fire Services Commission, 1998). This CIMS system was initiated in 1996 by the Fire Services of New Zealand and has its foundation
in the Incident Command System developed in Southern California in
1970 and the Australian Inter-service Incident Management System
(AIIMS) developed in the 1980s. CIMS is built around four major
components (NZ Fire Services Commission, 1998, p 14):

- CONTROL: management of the incident
- PLANNING/INTELLIGENCE: collection and analysis of in -
  cident information and planning of response activities

- OPERATIONS: direction of an agency's resources in combating the incident
- LOGISTICS: provision of facilities, services, and materials required to combat the incident.

External advice from scientific individuals, expert panels, agencies, and Science Advisory Groups (SAGs) or Committees (SACs) is commonly integrated into the Planning and Intelligence function described above, as well as being communicated directly to the Control decision maker.

Science advisory bodies have been called on during many volcanic crises worldwide, and best practice has identified that these advisory groups should be composed of many experts not only to pool expertise but also to combat issues that may arise due to conflict between scientists (Barclay et al., 2008). In NZ, many Science Advisory Groups have been formed over the last five years (see Smith, 2009), including the Central Plateau Volcanic Advisory Group (CPVAG) to advise officials about the Central Volcanoes in the North Island, the Auckland Volcanic Scientific Advisory Group (AVSAG) to advise officials about the volcanic field residing under Auckland City (discussed later), and the Tsunami Expert Panel, which forms in response to a local, regional, or distant, source earthquake and tsunami warning.

The advice provided by these technical and scientific experts is vital for the effective planning, intelligence gathering, and decision making of the emergency personnel and government officials in the protection of life, infrastructure, and welfare. However, to effectively provide advice to emergency managers, it is important to understand decision-making strategies and how this complex information is incorporated.

DECISION MAKING AND UNCERTAINTY

In Quarantelli's (1997) evaluation of the management of community disasters, the sixth criterion for good disaster management is to "permit the proper exercise of decision-making" (p. 46). From management decision-making research, a strategy for decision-making usually incorporates the stages (Flin, 1996, pp. 141–142):

1. "Identify the problem."
2. "Generate a set of options for solving the problem/choice alternatives."
3. "Evaluate these options concurrently using one of a number of strategies, such as weighting and comparing the relevant features of the options."
4. "Choose and implement the preferred option."

This can be considered to be an "analytic" decision-making process (Saaty, 2008). However, many decisions are made based on intuition, in a faster, almost automatic way (Flin, 1996). These decision-making processes have been studied extensively across a number of subdisciplines, including Classical Decision-Making (CDM), Behavioral Decision Theory (BDT), Judgment and Decision-Making (JDM), Organizational Decision-Making (ODM), and Naturalistic Decision-Making (NDM) (as reviewed in Lipshitz, Klein, Orasanu, & Salas, 2001). Decisions made in naturalistic settings have been characterized as involving (Orasanu & Connolly, 1993, as cited in Zsambok & Klein, 1997, p. 5):

1. "Ill-structured problems (not artificial, well-structured problems)."
2. "Uncertain, dynamic environments (not static, simulated situations)."
3. "Shifting, ill-defined, or competing goals (not clear and stable goals)."
4. "Action/feedback loops (not one-shot decisions)."
5. "Time stress (as opposed to ample time for tasks)."
6. "High stakes (not situations devoid of true consequences for the decision maker)."
7. "Multiple players (as opposed to individual decision making)."
8. "Organizational goals and norms (as opposed to decision making in a vacuum)."

Decision-making processes under these naturalistic conditions (NDM) can be defined as the way people use their experience to make decisions in real-world settings (Crichton & Flin, 2002; Klein, 2008; Zsambok & Klein, 1997). For critical incident management, research has identified four key NDM processes (Crego & Spinks, 1997; Crichton & Flin, 2002; Pascual & Henderson, 1997): (1) recognition-primed and intuition led action; (2) a course of action based on written or memorized procedures; (3) analytical comparison of different options for courses of action; and (4) creative designing of a novel course of action, ordered by increasing resource commitment.

The range of naturalistic decision-making models in the literature all consider that decision makers are utilizing and synthesizing prior experience to categorize situations and make decisions, rather than generating and comparing available decision options. The recognition-primed decision-making model of Klein (1998) considers that fireground commanders evaluate options without pattern matching or comparing to other options, instead using mental simulations to imagine the outcome of a decision and repeating until the first workable, rather than the best possible, outcome is found. The RPD model can thus be considered to involve a blend of both intuition and analysis (Klein, 2008). This model has evolved into three basic versions (see reviews in Flin, 1996; Klein, 1998; Lipshitz et al., 2001): (1) a simple match based decision based on the recognition of both a situation and the appropriate course of action; (2) a decision that requires diagnosis and assessment of the situation before a course of action can be implemented, often involving mentally simulating the events leading up to what is observed; and (3) a decision that requires an evaluation of the appropriate course of action often by mental simulation of the outcomes of those actions. Although these three levels may appear analytic, to the decision maker, they feel like a faster intuitive response (Flin, 1996).

A subset of NDM research concerns the study of emergency decision makers, where the particular pressures inherent in NDM are amplified—namely, the uncertainty, the high risk, the time pressure, and the constantly changing conditions. Martin, Flin, and Skriver (1997) identified that the dynamic decision-making processes occurring in these crises can vary considerably depending on the "task condition." In addition, different phases within any one incident may cause different decision-making processes to be adopted, ranging from the intuitive RPD approach through to the fully analytical approach, and the decision strategy may continue to switch as the situation continues to change. Martin et al. (1997) thus outline a framework for deci-

Cannon-Bowers et al. Task characteristic	*Intuition* *inducing* *(e.g. RPD)*		*Analysis* *inducing* *(e.g. option* *comparison)*
uncertain, dynamic	uncertain, fast	←——→	clear, slow moving
decision structure	ill defined	←——→	well defined
multiple goals	shifting/competing	←——→	clear/stable
time constraints	high (brief)	←——→	(low) long
quantity of	overload	←——→	adequate
decision complexity	low	←——→	high
multiple feedback loops	effect not yet determined		
risk/stakes	high	←——→	variable
multiple players	effect not yet determined		

Figure 5.1. Inducement of intuition and analysis in emergency situations as a function of task characteristics (conditions), after Martin et al. (1997, p. 283).

sion making in emergency situations that illustrates the inducement of an intuition or analysis approach-based on the emergency task "characteristic" outlined in Figure 5.1, suggesting that the emergency decision maker adapts their decision strategy depending on where the current situation characteristics sits on the various scales.

An emergency decision thus involves two distinct steps: (1) a definition of the problem and task characteristic via situation assessment (Endsley, 1997; Martin et al., 1997), and (2) a choice of what to do based on the four decision-making strategies (Crichton & Flin, 2002): (1) recognition-primed, (2) written or memorized procedures, (3) analytical, and (4) creative. Situation assessment (SA) addresses the question "What's the problem?" (Crichton & Flin, 2002, p. 209) and requires the correct evaluation of the event characteristics (Cannon-Bowers & Bell, 1997). It is considered by Endsley (1997, pp. 270–271) to have three levels:

- Level 1 SA: Perception of the elements in the environment in time and space.
- Level 2 SA: Comprehension of the current situation (in relation to the goals).
- Level 3 SA: Projection of the future status.

A decision maker may make the correct decision based on his or her perception of the situation, but if his or her situation assessment is incorrect, this may negatively influence his or her decision (Crichton & Flin, 2002). In making a decision, two factors are thus vital for the effective choice of action; the situation assessment and the *ongoing* situational awareness of the individual or team decision makers throughout the crisis (Sarna, 2002), as the different stages of the incident may require a different decision-making strategy and action.

Information Provision and Coping with Uncertainty

Information provision and science advice is a key component of the initial situation assessment and ongoing situation awareness during natural hazard crises. During the recent national CDEM Exercise Tangaroa conducted in NZ to test the response to a national tsunami warning, a threat level advisory map for the coast of New Zealand was provided by the science provider GNS Science and enabled rapid initial situation assessment by emergency managers (personal observations, October 20, 2010). For Exercise Ruaumoko, which tested the all-of-nation response to a volcanic eruption in the Auckland Volcanic Field, NZ (discussed further later; MCDEM, 2008a), ongoing situation assessment and awareness was improved by the information provided by the Auckland Volcanic Scientific Advisory group over a number of days. This ongoing advice allowed the emergency managers to adapt their decisions and responses accordingly as the situation evolved and the (theoretical) magma rose to the surface. In both of these exercises, the science advice was a key component of the emergency managers' situation awareness, influencing their action choices and the strategies used to make those decisions. However, as described by Harrald and Jefferson (2007, p. 3) "to provide the meaning component of situational awareness, the recipient of the information must be able to accurately perceive the situation described." Thus, it is not just the provision of the advice that is important, but also that the advice is correctly interpreted and meets the needs of the decision makers.

During a natural hazard crisis, this scientific information and advice may be sought by the emergency managers to address the uncertainty of the unfolding situation. Kuhlthau (1993; as cited in Sonnenwald & Pierce, 2000 p. 463) identified that "uncertainty is a cognitive state which causes anxiety and stress." Thus, any reduction of this uncertainty for emergency managers, via the provision of effective science advice, has the potential to reduce their anxiety and stress. After an extensive literature review of the terms *uncertainty, risk,* and *ambiguity* in decision making, Lipshitz and Strauss (1997, p. 150) define uncertainty in the context of action as "a sense of doubt that blocks or delays action" and identify that it can be "classified according to their issue (i.e., what the decision maker is uncertain about) and source (i.e., what causes this uncertainty)" (p. 151). From an analysis of decision makers' written accounts of dealing with uncertainty, they further identify that sources of uncertainty for action include (1) "incomplete information", (2) "inadequate understanding," and (3) "undifferentiated alternatives" (p. 151). Uncertainty in the context of the issue can relate to the outcome, the situation itself, and the alternative actions available. This uncertainty can occur on the level of the data, the level of the knowledge, and the level of understanding (Klein, 1998). Science advice may thus be subjected to uncertainties in the data due to (Patt & Dessai, 2005, pp. 426–427; van Asselt, 2000, p. 85): (1) the natural stochastic uncertainty, representing the random variability or chaotic nature of the system; and (2) the epistemic uncertainty, due to a lack of knowledge of the physical process.

Many formal and behavioral decision theories identify the R.Q.P. heuristic for coping with uncertainty in decision-making (see review in Lipshitz & Strauss, 1997), which represents the *reduction* of uncertainty by information searching, the *quantifying* of the magnitude of uncertainty that cannot be reduced, and the *plugging* of the result into a formal decision making scheme that incorporates uncertainty. However, this is limited by the need for multiple experts, the time to evaluate the uncertainties, and the issue that some uncertainties cannot be quantified into a numerical value (Lipshitz & Strauss, 1997). To investigate an alternative, Lipshitz and Strauss (1997) analyzed the written accounts of decisions made with uncertainty and found that the decision makers either:

- **Reduced the Uncertainty:** collecting additional information, deferring decisions until additional information became available, soliciting advice and following SOPs, and filling in their gaps in factual knowledge through assumption based reasoning
- **Acknowledged Uncertainty:** taking the uncertainty into account in the selection of an action by incorporating slack into the decisions/actions, improving readiness by generating new alternatives to pre-empt a specific potentially negative outcome, and weighing up the pros and cons of an approach.
- **Suppressed Uncertainty:** ignoring the uncertainty, relying on intuition, and rationalising and removing the doubts that block action. (p. 153-154, see also Lipshitz et al., 2001)

From this, they proposed their RAWFS (Reduce, Assumption-based reasoning, Weighing pros and cons, Forestalling, Suppressing) heuristic, stating that "how decision makers cope, or ought to cope, with uncertainty is principally determined by the nature or quality of the uncertainty" (Lipshitz & Strauss, 1997, p. 160). One of the key elements of the reduction phase of the RAWFS heuristic is the soliciting of advice and opinions of experts, demonstrating how science advice is not just about providing information for situation assessment but also about providing advice to help decision makers understand, acknowledge, reduce, or suppress the uncertainties in the source and the complex physical systems.

Team Decision Making: Shared Mental Models

Many decisions made in emergency management involve large and complex teams. In the NZ context, there are multiple team levels, from the team that operates within the Emergency Operation Center handling the response to the extended team that includes liaison officers from partner agencies. For the distributed decision making inherent in a multiorganizational emergency response such as this, people differ in their profession, expertise, functions, roles, and geographical location (Rogalski & Samurcay, 1993, as cited in Paton & Jackson, 2002), and thus the effective decision making of this distributed group is a function of the completeness of their shared mental model of the response environment in time and space, how their expertise contributes to different parts of the same plan, and their understanding of each others' knowledge, skills, roles, anticipated behavior or needs (Flin, 1996; Paton & Jackson, 2002).

If emergency managers turn to science advisors to solicit further information to guide their situation assessment and cope with uncertainty, then the communication between the science advisors and the emergency management team starts to play a vital role in the effectiveness of the entire decision process. However, it is not just a case of providing the emergency managers with all available science information but about understanding their needs to meet their information requirements. Simply providing as much advice as possible may actually hinder the decision process due to cognitive overload and an overuse of these available resources (Omodei, McLennan, Elliott, Wearing, & Clancy, 2005; Quarantelli, 1997). An on-site science advisor or an off-site expert panel can be considered to be part of the extended and distributed team handling the emergency management response. For a team to make effective decisions, many NDM concepts play a vital role, including team situation awareness, shared problem assessment, team mind, and shared mental models (see review in Lipshitz et al., 2001). To build up an understanding of the decision makers' needs and information requirements, the scientific advisors thus need to develop a shared mental model with the emergency managers. These shared mental models, or common knowledge bases, enable the team members to develop accurate expectations of the performance of themselves and their team mates and allow an effective coordination among team members without the need for extensive overt strategizing (Salas, Stout, & Cannon-Bowers, 1994).

A shared mental model is closely related to team situational awareness (SA), where team members have a specific set of SA elements about which they are concerned, determined by their individual responsibility (see review in Endsley, 1994). The overlap between each team member's SA then constitutes most of the interteam coordination (Endsley, 1994). If a team is composed of individuals with different levels of expertise, then the recognition of a situation or pattern will inherently come from an individual, while the interpretation of these recognized patterns is dependent on the conversational process of others in the team (Salas, Rosen, & DiazGranados, 2009). Orasnau (1994; as cited in Lipshitz et al., 2001, p. 341) identified that team SA can be achieved when team members "collect and exchange information earlier and plan farther in advance." Feedback between team members is then vital for the accuracy and refinement of these team member mental models (Salas et al., 1994). In the context of science advice for decision making, this supports the early integration of science information and involvement and a continued dialogue between the decision makers and advisors. By building a shared mental model, a team will have a shared understanding of the task, who is responsible for what, and what each other's information requirements are (Lipshitz et al., 2001).

Effective teams under high time pressure or increased workloads commonly adopt a communication style dominated by implicit supply of information rather than explicit requests, where members provide not only good information but unprompted information (see reviews in Kowalski-Trakofler, Vaught, & Scharf , 2003, p. 282; Paton & Jackson, 2002, p. 117). This implicit information gathering requires a good understanding from all team members of the information required by the main decision makers at critical periods. In addition, as discussed by Paton and Flin (1999, p. 261), if the format of this information does not require extensive additional information processing, while also matching the needs of the decision makers in terms of both content and format, then the stress for the decision maker can be reduced. Stress in a naturalistic setting may not cause poor decisions to be made based on the available information, but rather it can cause problems for information gathering, processing, and working memory (Crichton & Flin, 2002; Klein, 1998). Stress is thus likely to have a greater effect on decision-making styles that require generation and contrasting available options such as an analytical or a creative approach, rather than a recognition-primed or rule-based approach, due to the greater cognitive effort required for those strategies (Crichton & Flin, 2002; Klein, 1998).

Thus, as discussed by Paton (2003, p. 204), it is particularly important in these situations for the emergency communication from science advisors to involve (1) an anticipation and definition of information needs, (2) organized networks with information providers and recipients, and (3) an established capability to "provide, access, collate, interpret and disseminate information compatible with decision needs and systems." The science communication should tend more toward an implicit supply, through an understanding of the needs and demands on the decision maker, via a good team mental model that incorporates both the decision-making team and the science advisor. More multiorganizational and multidisciplinary planning activities with all team members and advisors will help in the development of similar mental models of the task (see review in Paton & Jackson, 2002), and collaborative exercises and simulations can further facilitate this understanding (Paton & Jackson, 2002). In addition, through the analysis of past events, lessons for successful communication, advice provisions, and distributed decision making can be learnt.

IMPROVING NZ'S RESPONSE CAPABILITY FROM 1995 TO 2007

The response to the eruptions at Ruapehu Volcano, NZ, from 1995 –1996 involved more than 40 agencies and organizations, with GNS Science acting as the major science provider (Johnston, Houghton, Neall, Ronan, & Paton, 2000). Analysis of the organizational response to these eruptions by Paton, Johnston, and Houghton (1998) identified that although 63 percent of the responding agencies reported getting information from GNS Science, limited formalized interorganization networking had been established prior to the event. This resulted in an ad hoc interaction with other agencies. As a result, science advisory processes in NZ have changed considerably since those eruptions. The use of a Science Advisory Group (SAG) representing the volcanological and social science expertise in an integrated emergency response was first fully tested during the recent National CDEM Exercise Ruaumoko. This occurred from November 2007 to March 2008 to test the local, regional, and national arrangements for dealing with the impact of a large natural hazard event on a major population center, and focused on the lead up to a volcanic eruption in the Auckland metropolitan area (MCDEM, 2008a).

Auckland sits on a "monogenetic" basalt volcanic field where eruptions occur over many different distributed volcanic vents. Over the last 250,000 years, 49 volcanic centers have been identified over the 360 km2 field, with the largest and youngest eruption approximately 600 years ago forming Rangitoto Island (see review in Lindsay et al., 2009). In the lead up to an eruption, felt earthquakes may cause a considerable amount of societal anxiety, as has occurred during seismic periods at other volcanic centres (e.g. Johnston et al., 2002). An eruption is likely to involve damaging earthquakes, explosions, and radially propagating base surges affecting buildings within 2 km of the vent and ash falling up to 10 km away (Lindsay et al., 2009; MCDEM, 2008a). As eruptions can occur anywhere within the Auckland Volcanic Field (AVF) and the location may not be known until magma is very close to the surface, decisions in exercises and any future real event will be typified by a high degree of uncertainty due to the eruption timing, location, severity, hazards, impacts, and consequences.

For Exercise Ruaumoko, the volcanic unrest scenario was developed in secret by a GNS Science volcano seismologist who did not participate in the exercise, and the exercise ended when the eruption started (see reviews in Lindsay et al., 2009; MCDEM, 2008a). Prior to the start of the exercise, the Auckland CDEM Group established and formalized the Auckland Volcanic Scientific Advisory Group (AVSAG), which contained a wide range of scientific expertise from GNS Science, NZ Universities (Auckland, Waikato, and Massey), and the Kestrel Group, as well as members of local and national CDEM (see reviews in MCDEM, 2008a; McDowell, 2008; Smith, 2009). AVSAG was conducted through a tripartite subgroup system (Monitoring, Volcanology, and Social), all of which reported upward to a smaller core SAG. During the days and weeks prior to the theoretical eruption, these subgroups liaised through teleconferences resulting in a coordinated advice provision to the NCMC and the Auckland Group EOC (AGEOC). This allowed for a direct question-and-answer dialogue to occur between the emergency managers and a wide range of science advisors. This dialogue supplemented the routine Scientific Alert Bulletins being produced on at least a daily basis by the GeoNet monitoring arm of GNS Science, and incorporated changes to the Volcanic Alert Levels for the AVF. Consistency between these two advice pathways was ensured by the fact that personnel writing the

GeoNet bulletins also were in AVSAG. In addition, as the scenario evolved, two on-site science advisors were dispatched from GNS Science to act as liaison officers within the NCMC and AGEOC to provide further advice and act as an information conduit among AVSAG, GeoNet, and the CDEM sector.

Postexercise reviews by MCDEM (2008a) and McDowell (2008) identified that this structure for science advice resulted in the science advice being very well delivered, clear, timely, and valuable. However, McDowell identified that during the most active periods of a volcanic eruption response, having separate subgroups composed of the predominately university-based "volcanology" group and the "monitoring" group of GNS-based scientists was unrealistic, and that the priority is not to have these separate groups but rather "the rapid assessment and decision making in relation to technical data" (McDowell 2008, p. 22). The presence of a science advisor in the Auckland CDEM Group EOC provided a "critical link for instant assessment and decision-making in relation to changing scientific information" (MCDEM, 2008a, p. 26), and other CDEM groups commented in their reviews that "the performance of the science providers has engendered a huge degree of confidence about their capacity and capability" (MCDEM, 2008a, p. 26).

MCDEM's (2008a) review further highlighted that the AVSAG approach's main strength lies in its inclusiveness of a wide range of scientific experts and competency, stating, however, that this inclusivity and "due process" slowed down the advice provision during the most active period. In addition, they identified the potential for a disconnect to occur between the local and national advice provision to the AGEOC and the NCMC, respectively. As highlighted by Cronin (2008), during the exercise, the two separate on-site advisors in these EOCs resulted in a divergence of the science advice as the event escalated, such that differences emerged in the evacuation planning at the local and national levels.

A potential disconnect may occur not only between the science advice to the local and national CDEM groups but also between the local and national science research *response, capability, and processes.* To address some of these limitations and coordinate the scientific advice beyond what is likely to be a limited knowledge pool and resources in a locally impacted area, an advisory group model is being coordinated by MCDEM (Smith, 2009), which will evolve to address the need for mobilization of NZ-wide science capability while remaining responsive to local CDEM needs. This model is described by Smith as having: "at its core national hazard monitoring capability and processes (e.g. GeoNet), with involvement of additional capability from universities and other science organisations based on thresholds of response. The intent is that GeoNet (both the technology and the science expertise of GNS Science) be the hub of any science response for earthquake, volcano, tsunami or landslide events" (p. 77).

This model still supports the existence of regions having existing scientific or planning advisory groups with a volcanic and/or earthquake focus, an example of which is the Central Plateau Advisory Group discussed earlier.

Beyond the successful coordination of the scientific response and the provision of science advice both nationally and locally, MCDEM (2008a) identified that communication between the advisors and the CDEM sector was negatively affected by the use of geological terminology, a degree of assumed knowledge, as well as challenges in interpreting scientific data. The scientific uncertainty of an event such as this, particularly with respect to the definition of the evacuation zones, the eruption timing, and impacts, creates a challenging environment for response planning and emergency management decisions. MCDEM (2008a) thus recommend that the frustration experienced by the CDEM sector due to this uncertainty would be reduced by an improved understanding of the AVF hazard, precursory signals, and a further translation of this primary science information to meet the needs of all organizations and limit misinterpretation.

ADVICE TAKING AND COMMUNICATING UNCERTAINTY

Existing research and case study reviews indicate that through collaborative preplanning, exercises, and training, which incorporates science advisory groups and bodies, the response capability and resilience of the entire integrated multiorganizational emergency response can be significantly enhanced. In developing these advice processes, relationships, and procedures, it is also important to consider the research fields concerned with advice taking, and the communication of uncertainty and probabilities.

Advice taking is considered by Harvey and Fischer (1997, p. 117) to be composed of three main components: "accepting help, improving judgement, and sharing responsibility," with the latter particularly apparent for experienced judges "when the risk associated with error was high." Whether this still applies in an emergency management decision-making context has not yet been investigated; however, the judgment literature has explored many aspects of advice taking and aggregation of opinions that are relevant in the CDEM context. This includes the degree to which people will take and utilize available advice, the role of multiple sources of advice, what happens when experts disagree, or there is conflicting advice, advice confidence, decision accuracy, and differences between advisors and decision makers (Bonaccio & Dalal, 2006; Budescu, Rantilla, Hsiu-Ting, & Tzur, 2003; Harvey & Fischer, 1997; Yaniv & Milyavsky, 2007).

Exercise Ruaumoko highlighted the benefits of science advice being provided by "one trusted source" during a crisis (MCDEM, 2008a), and forming this consensus requires a rationalization and integration of a wide range of potentially conflicting scientific opinions, model outputs, and outcome scenarios. Budescu et al. (2003) investigated the decisions made when a decision maker obtained probabilistic forecasts regarding the occurrence of a target event from a number of distinct, asymmetric advisors, where the asymmetry is due to the amount of information and the quality or accuracy of the advisors' previous forecasts. They found that the decision makers' final estimate can be described "as a weighted average of advisor forecasts, where the weights are sensitive to both [those] sources of asymmetry" (Budescu et al., 2003, p. 178). They identified that if the advisors' opinions were in perfect symmetry, then this was associated with the highest level of information fragmentation and the greatest degree of effort on the part of the decision maker who has to pay attention to all the opinions, reconciling the inconsistencies and disagreements. However, for a more asymmetric distribution of information, they postulate that it is easier for the decision maker to anchor his or her judgment to a smaller subset of the judges, and effort is reduced without sacrificing too much accuracy.

The importance of communicating a consensus opinion is further supported by the investigations of Yaniv and Milyavsky (2007), who found that when there was a wide range of advisor opinions, the decision makers revised their opinion in an egocentric manner, giving more weight to confirming opinions closer to their own than the disconfirming advice far from their own. However, this was sensitive to how much the decision maker already knew to begin with, with high-knowledge participants placing greater weight on their own opinion and on advice close to that, while the low-knowledge participants placed greater weight on the advice further from their own as well as utilizing all the advice more.

Due to the highly complex nature of volcanic eruptions, and the still many unknowns in volcano science, precise prediction is not achievable in many situations, and thus forecasts usually involve knowledge of both the dynamical phenomena and the uncertainties involved (Sparks, 2003). Thus, the use of numeric probability statements by scientists is becoming increasingly popular in volcanic crises due to a desire to make objective decisions via quantitative volcanic risk metrics and, ideally, predefined thresholds of probability based on a cost-benefit analysis (see review in Lindsay et al., 2009). In recent volcanic crises, consensus probability distributions have been produced via weighted and anonymous expert elicitation processes, which have then been fed into forecasting systems, such as Bayesian Event Trees (Aspinall & Cooke, 1998; Marzocchi & Woo, 2007). Adopting an approach such as this is highly advantageous for the decision-making process of the scientists, because it clarifies decision thresholds as well as optimizing the decision-making time (Lindsay et al., 2009). In addition, it offers the hindsight ability to clearly explain how a decision was made.

However, Solana, Kilburn, and Rolandi (2008) have also found that authorities at Vesuvius volcano, South Italy, prefer to receive deterministic statements instead of probability statements. Meanwhile, Haynes, Barclay, and Pidgeon (2008, p. 263) found that scientists at Montserrat Volcano Observatory, West Indies, considered using probabilities "to complicate communications as the likelihoods and associated uncertainties were neither well-explained nor understood." However, implicit in any evacuation decision by the authorities is the concern of making an "economically disastrous, unnecessary evacuation" (Tazieff, 1983; as cited in Woo, 2008, p. 88). Thus, authorities and other decision makers *do* commonly want to know the likelihood of potential hazard scenarios, likely casualty numbers, the effectiveness of an evacuation, and the uncertainties in the risk assessment (Woo, 2008) via a probabilistic decision theory methodology such as a cost-benefit analysis.

The weighted expert elicitation processes inherent in these volcanological decision aids help scientists to form a consensus and thereby reduce the effort required by emergency managers to "aggregate these opinions and generate a single response" when making a decision (Budescu et al., 2003, p. 178). However, once a collaborative, probabilistic communication is made, the probabilistic statements can commonly be misinterpreted. Research into the public understanding of probabilistic phrases has identified that the framing, directionality, and probabilistic format of these statements can bias people's understanding, affecting their action choices (e.g., Budescu, Broomell, & Por, 2009; Teigen & Brun, 1999). Thus, care must be taken with the presentation and wording of the communicated information, because it can affect the accuracy of the situation assessment made by individuals and teams and thus hinder the first crucial step in a decision-making process.

CONCLUDING REMARKS: BUILDING RESILIENCE THROUGH TRAINING

Relationship building before an event is vital for an effective response, as highlighted during the response to Hurricane Katrina, which Garnett and Kouzmin (2007, pp. 180–181) suggest was hindered by "differences in organizational culture and lack of trust that surfaced before Katrina had even formed." The effective communication across many agencies in a crisis is heavily dependent on this trust, as summarized by Comfort and Cahill (1998; as cited in Garnett & Kouzmin, 2007, p. 181), who state, "In environments of high uncertainty, this quality of interpersonal trust is essential for collective action. Building that trust in a multiorganizational operating environment is a complex process, perhaps the most difficult task in creating an emergency management system."

Group learning in a crisis context is thought to occur along three dimensions: personal, interpersonal, and institutional (Borodzicz & van Haperen, 2002). A number of research studies have investigated effective ways to train for naturalistic decision making (e.g., Cannon-

Bowers & Bell, 1997) and to enhance decision skills (e.g., Pliske, McCloskey, & Klein, 2001). Methods to train effective teams (e.g., Salas, Cannon-Bowers, & Johnston, 1997) have also been developed as part of research programs, such as TADMUS (Tactical Decision-Making Under Stress) (see review in Flin, 1996). Research has also identified methods to develop effective critical incident and team-based simulations (e.g. Crego & Spinks, 1997), where "in addition to knowledge and skill development, training should address how the disaster context influences performance and well-being" (Paton, Smith, & Violanti, 2000, p. 176). These simulations should also aim to reproduce reality as closely as possible, so decision makers and experts can experience the needs and realities of the advisory process in turbulent conditions (Borodzicz & van Haperen, 2002; Rosenthal & 't Hart, 1989). However, this alleged realism can also be a danger as personnel may believe at the end of an exercise that they know what will happen in a real crisis (Borodzicz & van Haperen, 2002). Evaluation of exercises and events must be carefully conducted to minimize the risk of creating an optimistic bias that overestimates future response preparedness and capability, particularly if a real event has not constituted a major test of the response system (Paton, Johnston, & Houghton, 1998).

In conclusion, the successful integration of science advice into the emergency decision-making process depends on a number of factors: (1) a well-functioning Emergency Operations Center with liaison officers from representative agencies and advisory bodies; (2) an environment and collaborative process that permits the proper exercise of decision making; (3) the experience and ability, or use of external experts, to develop an initial situation assessment and an ongoing situational awareness that permits an effective choice of action and decision style; (4) methods to reduce, acknowledge, or suppress uncertainty via the use of external experts and knowledge building about future potential crises; and (5) the development of relationships and a mutual understanding of each other's knowledge, skills, roles, and needs by both the emergency decision makers and their external advisors, promoting a shift from explicit requests to the implicit supply of science information. Only through an examination of historic case studies and participation in interorganizational simulations and training can these key response factors be developed to enhance future resilience and capability.

REFERENCES

Aspinall, W., & Cooke, R. (1998). Expert judgement and the Montserrat Volcano eruption. In A. Mosleh & R. A. Bari (Eds.), *Proceedings of the 4th International Conference on Probabilistic Safety Assessment and Management PSAM4, September 13th–18th* (Vol. 3, pp. 2113–2118). New York.

Asselt, M. B. A. (2000). *Perspectives on uncertainty and risk: The PRIMA approach to decision support.* Dordrecht, The Netherlands: Kluwer Academic Publishers.

Barclay, J., Haynes, K., Mitchell, T., Solana, C., Teeuw, R., & Darnell, A. (2008). Framing volcanic risk communication within disaster risk reduction: Finding ways for the social and physical sciences to work together. *Geological Society, London, Special Publications, 305*(1), 163–177.

Bonaccio, S., & Dalal, R. (2006). Advice taking and decision-making: An integrative literature review, and implications for the organizational sciences. *Organizational Behavior and Human Decision Processes, 101*(2), 127–151.

Borodzicz, E., & van Haperen, K. (2002). Individual and group learning in crisis simulations. *Journal of Contingencies and Crisis Management, 10*(3), 139–147.

Budescu, D. V., Broomell, S., & Por, H.-H. (2009). Improving communication of uncertainty in the reports of the intergovernmental panel on climate change. *Psychological Science, 20*(3), 299–308.

Budescu, D. V., Rantilla, A. K., Hsiu-Ting, Y., & Tzur, M. K. (2003). The effects of asymmetry among advisors on the aggregation of their opinions. *Organizational Behavior and Human Decision Processes, 90*(1), 178–194.

Cannon-Bowers, J. A., & Bell, H. E. (1997). Training decision makers for complex environments: Implications of the naturalistic decision-making perspective. In C. E. Zsambok & G. Klein (Eds.), *Naturalistic decision-making* (pp. 99–110). Mahwah, NJ: Lawrence Erlbaum Associates.

Crego, J., & Spinks, T. (1997). Critical incident management simulation. In R. Flin, E. Salas, M. Strub, & L. Martin (Eds.), *Decision-making under stress: Emerging themes and applications* (pp. 85–94). Aldershot, England: Ashgate Publishing Limited.

Crichton, M., & Flin, R. (2002). Command decision-making. In R. Flin & K. Arbuthnot (Eds.), *Incident command: Tales from the hot seat* (pp. 201–238). Aldershot, England: Ashgate Publishing Limited.

Cronin, S. J. (2008). The Auckland Volcano Scientific Advisory Group during Exercise Ruaumoko: Observations and recommendations data collection. In *Civil Defence Emergency Management: Exercise Ruaumoko.* Auckland, NZ: Auckland Regional Council.

Endsley, M. R. (1994). Situation awareness in dynamic human decision-making: theory. In R. D. Gilson, D. J. Garland, & J. M. Koonce (Eds.), *Situational awareness in complex systems: Proceedings of a Cahfa Conference* (pp. 27–58). Daytona Beach, FL: Embry-Riddle Aeronautical University Press.

Endsley, M. R. (1997). The role of situation awareness in naturalistic decision-making. In C. E. Zsambok & G. Klein (Eds.), *Naturalistic decision-making* (pp. 269–284). Mahwah, NJ: Lawrence Erlbaum Associates.

Flin, R. (1996). *Sitting in the hot seat: Leaders and teams for critical incident management.* Chichester, England: John Wiley & Sons, Ltd.

Garnett, J. L., & Kouzmin, A. (2007). Communicating throughout Katrina: Competing and complementary conceptual lenses on crisis communication. *Public Administration Review: Administrative Failure in the Wake of Katrina, 67*(1), 171–188.

Harrald, J., & Jefferson, T. (2007). Shared situational awareness in emergency management mitigation and response. *Proceedings of the Fortieth Annual Hawaii International Conference on System Sciences (HICSS'07), Waikoloa, HI,* 23-23. Los Alamitos, CA: IEEE Computer Society.

Harvey, N., & Fischer, I. (1997). Taking advice: Accepting help, improving judgment and sharing responsibility. *Organizational Behavior and Human Decision Processes, 70*(2), 117–133.

Haynes, K., Barclay, J., & Pidgeon, N. (2008). Whose reality counts? Factors affecting the perception of volcanic risk. *Journal of Volcanology and Geothermal Research, 172*(3–4), 259–272.

Johnston, D. M., Houghton, B. F., Neall, V. E., Ronan, K. R., & Paton, D. (2000). Impacts of the 1945 and 1995–1996 Ruaupehu eruptions, New Zealand: An example of increasing societal vulnerability. *Geological Society of America Bulletin, 112*(5), 720–726.

Johnston, D. M., Scott, B., Houghton, B. F., Paton, D., Dowrick, D., & Villamor, P. (2002). Social and economic consequences of historic Caldera unrest at the Taupo Volcano, New Zealand and the management of future episodes of Unrest. *Bulletin of The New Zealand Society For Earthquake Engineering, 35*(4), 215–230.

Klein, G. (1998). *Sources of power: How people make decisions* (2nd ed.). Cambridge, MA; London, UK: MIT Press.

Klein, G. (2008). Naturalistic decision-making. *Human Factors: The Journal of the Human Factors and Ergonomics Society, 50*(3), 456-460.

Kowalski-Trakofler, K. M., Vaught, C., & Scharf, T. (2003). Judgment and decision-making under stress: An overview for emergency managers. *International Journal of Emergency Management, 1*(3), 278–289.

Lee, B. Y. (2010). Working together, building capacity–A case study of civil defence emergency management in New Zealand. *Journal of Disaster Research, 5*(5), 565--576.

Lindsay, J., Marzocchi, W., Jolly, G., Constantinescu, R., Selva, J., & Sandri, L. (2009). Towards real-time eruption forecasting in the Auckland Volcanic Field: Application of BET_EF during the NZ National Disaster Exercise "Ruaumoko." *Bulletin of Volcanology, 72*(2), 185–204.

Lipshitz, R., Klein, G., Orasanu, J., & Salas, E. (2001). Focus article: Taking stock of naturalistic decision-making. *Journal of Behavioral Decision-Making, 14*(5), 331–352.

Lipshitz, R., & Strauss, O. (1997). Coping with uncertainty: A naturalistic decision-making analysis,. *Organizational Behavior and Human Decision Processes, 69*(2), 149–163.

Martin, L., Flin, R., & Skriver, J. (1997). Emergency decision-making–A wider decision framework? In R. Flin, E. Salas, M. Strub, & L. Martin (Eds.), *Decision-making under stress: Emerging themes and applications* (pp. 280–290). Aldershot, England: Ashgate Publishing Limited.

Marzocchi, W., & Woo, G. (2007). Probabilistic eruption forecasting and the call for an evacuation. *Geophysical Research Letters, 34*(22), 1–4.

MCDEM. (2008a). *Exercise Ruaumoko '08 Final Exercise Report.* Wellington, NZ: Department of Internal Affairs.

MCDEM. (2008b). *National Civil Defence Emergency Management Strategy 2007.* Wellington, NZ: Department of Internal Affairs.

McDowell, S. (2008). *Exercise Ruaumoko: Evaluation Report Auckland Civil Defence Emergency Management Group.* Auckland, NZ: Auckland Regional Council.

NZ Fire Services Commission. (1998). *The New Zealand Coordinated Incident Management System (CIMS): Teamwork in Emergency Management. Management* (1st ed.). Wellington, NZ: Author.

Omodei, M. M., McLennan, J., Elliott, G. C., Wearing, A. J., & Clancy, J. M. (2005). "More is better?": A bias toward overuse of resources in naturalistic decision-making settings. In H. Montgomery, R. Lipshitz, & B. Brehmer (Eds.), *How professionals make decisions* (pp. 29–41). Mahwah, NJ: Lawrence Erlbaum Associates.

Pascual, R., & Henderson, S. (1997). Evidence of naturalistic decision-making in military command and control. In C. E. Zsambok & G. Klein (Eds.), *Naturalistic decision-making* (pp. 217–226). Mahwah, NJ: Lawrence Erlbaum Associates.

Paton, D. (2003). Stress in disaster response: A risk management approach. *Disaster Prevention and Management, 12*(3), 203–209.

Paton, D., & Jackson, D. M. (2002). Developing disaster management capability: An assessment centre approach. *Disaster Prevention and Management, 11*(2), 115–122.

Paton, D., Johnston, D. M., & Houghton, B. F. (1998). Organisational response to a volcanic eruption. *Disaster Prevention and Management, 7*(1), 5–13.

Paton, D., Smith, L., & Violanti, J. (2000). Disaster response: Risk, vulnerability and resilience. *Disaster Prevention and Management, 9*(3), 173–180.

Patt, A., & Dessai, S. (2005). Communicating uncertainty: Lessons learned and suggestions for climate change assessment. *Comptes Rendus Geosciences, 337*(4), 425–441.

Pliske, R. M., McCloskey, M. J., & Klein, G. (2001). Decision skills training: facilitating learning from experience. In E. Salas & G. Klein (Eds.), *Linking expertise and naturalistic decision-making* (pp. 37–53). Mahwah, NJ: Lawrence Erlbaum Associates.

Quarantelli, E. L. (1997). Ten criteria for evaluating the management of community disasters. *Disasters, 21*(1), 39–56.

Rosenthal, U., & 't Hart, P. (1989). Managing terrorism: The south Moluccan hostage takings. In U. Rosenthal, M. T. Charles, & P. 't Hart (Eds.), *Coping with crises: The management of disasters, riots and terrorism* (pp. 340–366). Springfield, IL: Charles C Thomas.

Saaty, T. L. (2008). Decision-making with the analytic hierarchy process. *International Journal of Services Sciences, 1*(1), 83–98.

Salas, E., Cannon-Bowers, J. A., & Johnston, J. H. (1997). How can you turn a team of experts into an expert team? Emerging training strategies. In C. E. Zsambok & G. Klein (Eds.), *Naturalistic decision-making* (pp. 359–370). Mahwah, NJ: Lawrence Erlbaum Associates.

Salas, E., Rosen, M. A., & DiazGranados, D. (2009). Expertise-based intuition and decision-making in organizations. *Journal of Management, 36*(4), 941–973.

Salas, E., Stout, R. J., & Cannon-Bowers, J. A. (1994). The role of shared mental models in developing shared situational awareness. In R. D. Gilson, D. J. Garland, & J. M. Koonce (Eds.), *Situational awareness in complex systems: Proceedings of a Cahfa Conference* (pp. 298–304). Daytona Beach, FL: Embry-Riddle Aeronautical University Press.

Sarna, P. (2002). Managing the spike: The command perspective in critical incidents. In R. Flin & K. Arbuthnot (Eds.), *Incident command: Tales from the hot seat* (pp. 32–57). Aldershot, England: Ashgate Publishing Limited.

Smith, R. (2009). Research, science and emergency management: Partnering for resilience. In *Tephra, Community Resilience: Research, Planning and Civil Defence Emergency Management* (pp. 71–78). Wellington, NZ: Ministry of Civil Defence & Emergency Management.

Solana, M. C., Kilburn, C. R. J., & Rolandi, G. (2008). Communicating eruption and hazard forecasts on Vesuvius, Southern Italy. *Journal of Volcanology and Geothermal Research, 172*(3-4), 308–314.

Sonnenwald, D. H., & Pierce, L. G. (2000). Information behavior in dynamic group work contexts: Interwoven situational awareness, dense social networks and contested collaboration in command and control. *Information Processing & Management, 36*(3), 461–479.

Sparks, R. S. J. (2003). Forecasting volcanic eruptions. *Earth and Planetary Science Letters, 210*(1–2), 1–15.

Teigen, K. H., & Brun, W. (1999). The directionality of verbal probability expressions: Effects on decisions, predictions, and probabilistic reasoning. *Organizational Behavior and Human Decision Processes, 80*(2), 155–190.

Woo, G. (2008). Probabilistic criteria for volcano evacuation decisions. *Natural Hazards, 45*(1), 87–97.

Wood, P., Robins, P., & Hare, J. (2010). Preliminary observations of the 2010 Darfield (Canterbury) earthquakes: An introduction. *Bulletin of the New Zealand Society for Earthquake Engineering, 43*(4), i–iv.

Yaniv, I., & Milyavsky, M. (2007). Using advice from multiple sources to revise and improve judgments. *Organizational Behavior and Human Decision Processes, 103*(1), 104–120.

Zsambok, C. E., & Klein, G. (Eds.). (1997). *Naturalistic decision-making*. Mahwah, NJ: Lawrence Erlbaum Associates.

Chapter 6

COP SHOT

JAMES J. DRYLIE

INTRODUCTION

In an effort to better understand resiliency and how someone can maintain his or her existence in the face of extreme odds, an exhaustive case study analysis was conducted of a police officer who was shot in the line of duty during an unprovoked attack. The officer's story is unique, not remarkable. His life growing up was hard, not impossible. The ability of the officer to recover and return to work in the face of overwhelming odds did not occur in a vacuum, it did not happen overnight. The story of Ken Hogan is a story of resiliency. This chapter offers a glimpse into his life before, during, and after a shooting that left him near death, slumped in the front of a marked radio car with a combined total of 19 rounds in both his body and the police car.

SEEDS OF RESILIENCY

There are many iconic images associated with the resiliency of the American people. The raising of the American flag at Iwo Jima and Ground Zero are symbolic of a nation willing and capable to meet and overcome adversity. In the wake of disaster after disaster, these visual images of strength, unity, and hope undoubtedly capture the hearts and minds of Americans, serving as a reminder of the spirit that has made this nation great.

Other images associated with resiliency are more subtle but profound nonetheless. Perhaps you have seen photographs of landscapes that were devastated by fire, flood, or drought, and in the midst of the destruction is a sign. A glimmer of hope. In what seems to be a barren wasteland, there are signs of life. The emergence of a budding flower or seedling demonstrates resiliency in the face of extreme odds against survival. Yet this renewed growth began before the incident resulting in the vast destruction of life. Using this image as a metaphor for human resiliency makes sense when you consider that the seeds that result in this resurgence were in place before disaster struck.

This narrative is an examination of human resiliency that parallels that of the seedling or budding flower that grows despite of destructive forces. Based on a case study that focused on an unprovoked assault of an on-duty uniformed officer, this narrative exposes the harshness of life experienced by one man in the face of personal adversity, traumatic injury, and a deep personal resolve never to give up.

The Ken Hogan story is one of resiliency, a resiliency that developed very early in a young man's life. His childhood was clouded with pain. When he needed profound physical and emotional strength to survive from multiple gunshot wounds, Ken's resiliency, which is a product of both good and bad lifelong experiences, helped him to follow a plan. Ken Hogan had a plan—that plan was to live.

Ken's story, like so many others, began before he is born. Similar to the seed, the odds were against him from the beginning. The youngest of four children, Ken was born in Irvington, New Jersey, in the summer of 1958. His father, according to family accounts, was extremely angry with his mother over this new addition to the family. Ken was just another mouth to feed. Both of his parents were alcoholics, and his father was incapable of maintaining steady employment, and his mother would ultimately lose her job as well due to her drinking. What money was earned was often consumed by the parents' addiction to alcohol and gambling.

In the early years, Ken's family lived in an industrial area of the city. His father was absent more than he was home. Ken's father would not have a chance to leave a positive impression on the young Hogan. When he was just three years old, Ken's father was savagely beaten near their home. When the father came into the house, battered and broken, he was cared for by Ken's sister; their mother slept through the night unaware of what had happened. The following day, his father was taken by ambulance to Irvington General, the same hospital where Ken was born. He died four days later from his injuries. The murder was likely the work of elements of organized crime connected with his father's gambling debts. Death had entered young Ken Hogan's life.

After his father's murder, life for Hogan and his family would continue to be a struggle, but they were not without help and support from the local community. St. Leo's, the neighborhood Catholic church, offered assistance to Ken's mother by helping with rent and allowing the children to attend parochial school. Sister Maureen Williams was Ken's first-grade teacher at St. Leo's. He recalled the nun telling him to be strong and never to be afraid. Her words would not be lost on the young six-year-old. Sister Maureen had a profound impact on young Hogan, and he would be devastated the following year when the nun died from leukemia. Death had visited again.

After two more years at Catholic school, Ken's personal and family struggles would persist. He was taken out of St. Leo's and enrolled in a local public elementary school. Living in federal housing, early memories of his home life were overshadowed by his mother's drinking. Ken remembers having to go to the local tavern for his mother, and after passing an IOU to the bartender, he would return home with a pint of draft beer in a *tainer,* slang for the heavy white paper container. Those times when the beer spilled, or worse yet the bag ripped, Ken would suffer from his mother's indignation.

As a young adolescent, Ken's life would provide a steady diet of abuse, neglect, and disappointments. People who were expected to be heroes to a young boy were not. Promises were more often than not broken, and family efficacy was virtually nonexistent. The mounting negativity became a positive force in Hogan's life. He was going to survive. He was going to make a difference. It was his plan.

Ken Hogan was a young man when he joined the police force in Irvington, New Jersey. Hogan learned early in his career, as a young recruit in the Essex County Police Academy, of the harsh realities of life when a veteran Irvington police officer was killed in a motor vehicle accident. The murder of New Jersey state trooper Phillip LaMonaco in December 1981 by radical revolutionaries also deeply affected Hogan. Like so many police officers, he wanted to be a part of the effort to arrest those responsible for the death of the highly decorated state trooper. More than this, Hogan began the process of mental preparation as a means of self-preservation. The beginning of Ken Hogan's plan for survival was taking shape. The murder of Tony Graffa, an Irvington police officer, had a profound effect on Hogan. Officer Graffa was a cop to emulate. Ken looked to him as "a cops' cop." Unlike the people in young Hogan's life who let him down, failed to keep promises, or were just not there, Tony Graffa was a true hero. Hogan cried for days after Graffa's death.

In just five short years as a police officer, Ken Hogan knew of too many officers who died in the line of duty, and equally as disturbing were cases involving police suicide. Although he did not expect to live forever, life was truly too short. Hogan's plan was to keep the playing field level by sticking to his plan—to survive.

Nothing Is Routine

The ride into work was different today. It was a Monday, late January 1994, and the ground was covered from a recent snowfall. Ken Hogan was making the trip—a trip he made every workday, into the City of Irvington, New Jersey, a struggling urban community bordering Newark, the state's largest city. A 16-year veteran of the Irvington Police Department, Ken was a highly decorated and a seasoned officer, factors that would soon play a pivotal part in his plan to survive later that morning as he lay gravely wounded from multiple gunshot wounds.

The commute for Hogan was a relatively short distance, but commuting in northern New Jersey during rush hour is an endeavor that requires both physical and mental stamina. Ken was thinking about his upcoming wedding. The day before, Sunday, family and friends hosted a wedding shower for Ken and his fiancée at *Cryan's Ale House,* a popular Irish pub in South Orange. The ride seemed quicker than usual, and as Ken passed through the pastoral suburban landscape, the blanket of white snow along the roadway soon gave way to a dull shade of gray. The purity of the fresh snow symbolized the innocence of better times. The visual transformation from white to gray was much more than the natural occurrence of snow removal efforts, traffic, and pollution; it represented the loss of innocence that cities like

Irvington suffered in the later part of the twentieth century. It is in this environment that officers like Ken Hogan worked and in some cases died.

Roll call was roll call. The banter between the cops was the same—talk focused on the past weekend and plans for the upcoming week. Coffee was a beverage of necessity for many. Light and sweet, black with no sugar, or just milk; the preferences were as diverse as the faces in the room. Hogan was no different; he had his likes and dislikes. He glanced around the room and identified cops as belonging to two categories: good cops and not so good cops. On this Monday morning, there was a little of both present, but he didn't dwell on it. He continued to mentally prepare for the day.

If his radio car assignment, Car #8, was any indication of how his day was going to be, Ken was in for a long day. The car, a Chevrolet Impala, had more than 80,000 miles and showed all the signs of wear; a real eyesore. Ken preferred the older cars, particularly Chevy's. Irvington was a city, streets were narrow, and operating a police car required a vehicle that the cops could count on. In his opinion, the steering and transmissions on the Chevy's handled better. Hogan had a routine of inspecting the radio car inside and out. Today was no different. The first thing he would check before even placing his gear into the car would be the backseat. Hogan looked for everything, anything that could possibly be there from the previous shift. Guns, drugs, ammunition, whatever could be stuffed in the backseat during holding and transport of suspects, he would check for it. His vehicle check routine would include the trunk area as well. Ken wanted to ensure that the vehicle was equipped with a spare tire, First-Aid supplies, and flares. He also checked to see whether any personal gear from other officers was left behind. Cops always leave personal shit in the trunk, some things more personal than others. After completing a check of the passenger and trunk areas, Hogan popped the hood and looked for obvious indicators of mechanical problems like oil leaks, excess fluid, or broken hoses. The final check was of the vehicle's exterior and tires. Dents, scratches, and obvious signs of wear were common. Many of the cars showed significant wear on the outside, but Hogan was looking for recent damage that required reporting to his supervisor and the condition of the tires. An admitted hard driver, Hogan knew that adequate tire pressure and tread were critical to good operation of the car under extreme conditions. Satisfied with the overall inspection, Hogan started the car.

The engine is missing. That's just great.

Hogan left headquarters and drove straight to the City Motor Pool on 16th Avenue. At 7:00 a.m. the garage was not yet officially open, but he was able to get a couple of mechanics to help him check the car out. Ken had always maintained a good relationship with the mechanics. In the back of his mind, he knew that he would and could rely on that relationship in a pinch. This was one of those times.

After several attempts to correct the engine miss, they replaced spark plugs and wires. The decision to deadline the car was made, and a substitute car was assigned from the motor pool. Hagan repeated the inspection of the newly assigned car, #7, and was ready to begin his tour, with one exception. The mechanic did not have a single spare key for Car #7, so the garage foreman handed him a ring that held more than 50 car keys. This huge key ring was not something that Hogan or any officer could place in their pants pockets or on a gun belt. As a matter of routine, Hogan would leave keys on the floor when he needed to exit the car quickly. His thoughts focused on how he would do that with a key ring that a school janitor would envy.

What am I suppose to do with 50 keys? One key, that's all I need, one key. I got 50, 50 keys. Only in Irvington. What a job.

After clearing from the municipal garage, Hogan began his way toward his assigned area: 203. The city was divided into four areas, each about the same size in terms of square miles covered. Many of the homes in the area were two-family, neatly lining the streets side by side.

A native of the city, Hogan first worked 203 as a beat cop. Long before community policing was in vogue, Hogan would be invited by area residents to attend birthday parties and enjoy some home-cooked meals during local barbeques. Most area residents were black, and second- or third-generation Irvington residents. But things changed. Many of the people that Hogan knew as a young beat cop were gone. Many homes were vacant, some boarded up, others just abandoned. There were homes that stood out, homes that were cared for. Sadly, many of these properties were ringed with heavy gates and fences in attempts to hold back the crushing advance of urban decay and blight. There was hope. There were people who cared; there were people like Ken Hogan.

As he made his way east on Springfield Avenue into the 203 area, Hogan's thoughts drifted between the present and the future. Thoughts of an upcoming promotion to the rank of sergeant were mixed with practical thoughts of how he would help to prepare and protect the cops who would eventually work for him. Some of the cops were good and some not so good. The not so good ones were his main concern. Cops can get into trouble. They get hurt. They die. Not that day, not Ken Hogan; he has thought a lot about this, and that is not a part of his plan.

Turning off of Springfield Avenue onto 21st Street, Officer Ken Hogan began to clear his mind and focused on the area and hidden dangers. Hogan was keenly aware that the streets are often used as a cut-through by car thiefs from the area to travel into and out of the City of Newark. The area is a classic example of social apathy. Abandoned houses line the streets, and his conscious thoughts are about survival–his own and that of the people in the neighborhood.

Traveling north on 21st Street, Hogan approached the intersection of Nelson Place and Standard Place, a typical residential street corner in the City of Irvington. He continued north into a one-way street. The absence of cars made his choice of driving against the flow of traffic easier. If anyone was out on the street, the element of surprise would be in his favor.

As he drove slowly down the street, one of the first things that Hogan noticed was out of place was something that usually is present, but it's not: garbage. Normally garbage would be along the street and collected near the curbs. The snow was hiding all of that, at least for the time being. It will come back, it always did. Something else was missing: people. No one was out. The porches were empty. He saw some kids looking out from a window; he recognized friendly faces, local kids. Inside, warm, a good place for them on a day like this. Hogan reflected on how peaceful the streets were at that moment. That was all about to change.

As Hogan approached the intersection of 19th Avenue and 21st Street, he glanced toward the east, toward Newark. Nineteenth Avenue is a clear shot into Newark, and Hogan's thoughts was focused on stolen motor vehicles. The street was clear. At that moment, Hogan noticed a black male walking in a westerly direction toward his marked radio car. The male was on the sidewalk as he approached Hogan.

Hogan didn't recognize this guy. His first thought was that he was not from the area. There are no bus stops, no stores, and no schools. What was he up to? Hogan was in *cop mode*. His radar was on. This is a drug area; people buy and sell drugs whether there is snow on the ground or not. Stolen car? What's the deal?

As the male approached, Hogan keyed in on his body language. The male was looking down. Hogan wondered why he was not looking up at the radio car. The only moving car on the street, the radio car must be obvious. As he assessed the situation, Hogan noticed that the male continued to look down, avoiding any eye contact or acknowledging that the police car was even there.

What's with this guy? Something's not right. He's not looking at me. Why isn't he looking at me? Come on Mr. Happy, what's up?

As he continued to assess the male and his body language, Hogan's instincts told him that he was likely spotted by this guy first. The male hadn't done anything wrong, yet. Hogan's observations were part intuition, part survival. As he continued to read the man's body language, Hogan noticed that the guy's arms stiffened and his direction on the sidewalk began to move from right to left. The guy began walking from the side of the sidewalk closest to the street to the inside of the sidewalk, closer to the houses.

This guy was not playing by any script that Hogan expected. Hogan began to search his memory for past events that had occurred in this area. A recent gun job, a car chase, or drug deals were thoughts that quickly came to mind. As the male approached the intersection, Hogan inched the police car farther toward the sidewalk in an attempt to gain the man's attention. The man's pace did not change or slow down at all.

This ain't right, his body is too stiff. What is this guy up to?

Hogan was still in the police car as the male suddenly changed direction and began walking toward the rear of the radio car. Hogan was forced to glance over his right shoulder, looking across the interior of the radio car. As the male continued past the police car, Hogan felt a sense of vulnerability as he was now forced to turn from his right to his left to continue to observe the man.

As Hogan continued to watch the male, he expected him to cross back toward the crosswalk area to continue walking west. Instead, he walked southeast toward a house that Hogan was familiar with from past incidents. Hogan had made numerous arrests in the area in the past, including several gun arrests, and based on his instincts believed that the male was possibly looking to make a buy.

There were two males standing on a porch adjacent to where Hogan last saw the male walking. Neither of these men was present when Hogan first drove past the house. Hogan's suspicions were mounting, and he began to view the male in a different light: suspect. Hogan's state of awareness intensified. He scanned the neighborhood looking for escape routes in case this guy took off. Hogan's ready for action.

Hogan continued in the same direction he was traveling when he first observed the lone male. He decided to turn the radio car around, to bring things back into his terms. After a quick K-turn, Hogan was again in front of the house where moments before he observed the two males on the front porch.

Where did they go? The two guys, they're gone. Where are they?
Mr. Happy's moving fast.

The two males on the porch were gone. Likely back in the house, but the suspect was moving away; he was approximately 75 feet away from the house heading back toward the direction where Hogan first observed him. Hogan continued driving the radio car in the direction of the suspect. He remembered the two on the porch; they were local street-level drug dealers, he was certain of it.

This guy's not fazed by me. Look he's tightening up. He's up to no good.

Hogan was mentally preparing himself for the stop. He scanned his memory for past arrests in the area. He checked for familiar signs and cues that could help in the event that there was any type of foot pursuit.

He's gonna do something. Rabbit, fight, something, he's got to.
This guy just bought or sold I know it.

The suspect continued to walk away from Hogan, but he was headed toward the dead end section of Nelson Place. Hogan moved the car in the direction of the suspect. The sound of the radio car would have been obvious to anyone walking in the area, but this guy did not react.

He's dirty, he's got something on him, I can feel it.

As Hogan continued toward the suspect, he positioned the radio car in such a way that he could turn in any direction to chase the suspect on foot or with the car if he took off. Hogan was feeling confident. He made the conscious decision to stop the suspect. With a sense of confidence he radioed to headquarters:

> **Hogan:**[1] *Gonna be stopping one on 21st and Nelson.*
> **Headquarters (HQ):** *Alright. Unit in the area of Nelson and 21st, to back him.*

It was standard procedure in the Irvington Police Department to dispatch a back-up unit to assist any officer during a field interview or motor vehicle stop. Hogan was aware of this and did not acknowledge that back-up was assigned. There was nothing inherent in Hogan's radio transmission or his voice that alerted communications or other personnel that the officer was in imminent danger. He had made that same transmission hundreds of times in his career. He was a skilled, veteran cop. Ken Hogan had a plan, and he was acting according to the script he had played over and over in his mind.

The suspect's back was to Hogan at this point. Hogan was watching him for any sign or indication of what would happen. Hogan recalled that the suspect's jacket was black, but his attention quickly moved toward his hands.

> *Alright Mr. Happy, what's in your hands, anything? Do you have anything in your hands?*

Hogan was searching for something, anything that would indicate what the suspect's intentions were. His hands were open; they weren't clenched or closed in any way that would be an attempt to hide something, anything. He wasn't holding a damn thing; money, drugs, anything, but his arms, they were rigid, stiff. His body language was sending signals to Hogan, loud signals.

> *What's he up to? Mr. Happy, you're up to something, yes you are, I can feel it. Watch this guy Kenny, he's up to no good.*

Hogan had stopped individuals in similar situations countless times, hundreds of times, but he wasn't taking anything for granted. He ran over in his mind what his actions would be, what he would say. He prepared himself like he always did. What if he runs? Hogan was

---

1. Radio transmissions are between Hogan and the Irvington Police Department Communications Center.

beginning to process how he would broadcast the information. He continued to assess his location. Mentally he established the compass points: North, South, East, and West. He knew that it was critical to give accurate locations. Hogan's senses were heightened. His mind processed both conscious and subconscious thoughts. The suspect and his actions were in the forefront of Hogan's mind. His subconscious was handling the situation based on 16 years of police work, 16 years of surviving. Hogan was a survivor. He sat erect in the car, his eyes taking in everything; the suspect, the neighborhood, likely escape routes. Hogan was in full *cop mode.*

The suspect was about 35 feet from the front of the police car. Hogan was prepared to act. He continued to assess the situation. His eyes were scanning every possible inch of the suspect's body. The suspect's body was still stiff. Hogan could clearly see his back and his arms. Clear, they were clear, nothing visible. The driver's side window of the police car was down all the way. Hogan began to open the door and calls out to the suspect.

Hogan: *"Yo, my man, hold up a minute."*

As he called out to the suspect, Ken noticed that the suspect began to turn slightly to his right. At this point, Hogan lost sight of the suspect's right hand.

Where is his hand? I can't see his hand. I gotta see his hand.

The suspect began to move. Hogan was following the movement. Things began to happen rapidly, quicker than Ken would have expected. The suspect's jacket, a black army-type jacket, was open. The movement concealed his right hand. Hogan could no longer see the suspect's right hand.

His hand, where is his hand? I can't see his hand. This ain't good, I can't see his hand. GUN.

Gun. It was as clear as if this were designed to happen. The suspect had removed a handgun from his waistband.

Gun. He's got a gun.

The suspect swung his right arm up pointing the gun in the direction of Hogan's police car. Hogan saw a flash, there was a loud crack.

POP.

Everything was bright. Hogan recalled that the sun was reflected off the snow that was on the ground. Hogan's senses were at peak levels; he was processing things that seemed insignificant.

The snow, the snow.

He was processing everything that had happened. Checking and cross-checking. He was keenly aware of his surroundings. Hogan was still in the car; he never exited it.

MOVE HOGAN, MOVE. MOVE. GET DOWN, GET DOWN.

Hogan moved instinctively to his right, it was his only option. The first shot hit the windshield at a point even with Hogan's head. The round was dead on. The interior of the police car filled with millions of particles of glass. The glass moved toward Hogan's face.

The windshield, the windshield is hit.

Hogan's physical reactions to the initial gunshot were going as planned. He had planned for this moment his whole career. The car would provide cover. Prepare for the pain. Use the mind to control the body. Hogan was preparing for the inevitable.

As quickly as everything was happening, it seemed as if time had slowed to a crawl. They told him this would happen. It's happened before. Hogan knew the script, and he was living it. Hogan's movements were quick, reactive to the danger, yet each movement that he made seemed to take forever. The sense of slow motion was magnified, it even seemed normal.

Hogan continued to move to his right, first hitting the armrest between the front seats.

Move, get down, down. Don't let him get you. Move.

Hogan's head was down. The passenger seat prevented further downward movement. Was he trapped? Would the design of the car be his downfall? No. He continued to reduce the chance of exposure. Moving toward the passenger side door Hogan heard several more gunshots.

POP. POP. POP. POP. POP. An auto, this guy's got an automatic. How many rounds did he fire? What is it, what kind of gun is it?

HE'S COMING. THIS GUY IS COMING FOR ME.

As Hogan continued to move toward the passenger side of the patrol car, he realized that the suspect was not stopping; this guy was on a mission. His mission was to kill a cop. His mission was to kill Ken Hogan. Hogan's plan was to stop him–stop him from completing his mission; Hogan's mission was to survive.

Oh my God. Oh my God.

The suspect reached the driver's side door.

How did he get here? This can't be happening? How did he get here?

Hogan's thoughts were on escape. There was no time for anything else.

Get out of the car. Get out of the car Hogan. MOVE. MOVE. MOVE. Move Kenny, you have to get the hell out of here.

Hogan heard several more gunshots, but by then he had lost count, the rounds were fired without hesitation.

Did he reload, is he out of ammo? He ain't stopping.

Hogan was in full survival mode. His thoughts focused on his escape. The door, he needed to move toward the door.

Get out of the car. You're trapped, get out of the car. DO IT NOW HOGAN, DO IT NOW, GET OUT OF THE CAR.

Hogan reached for the passenger side door handle with his right hand. He had rehearsed this very scenario in his mind several times before, over and over again. These were the "What if" scenarios. This is what he had been trained to do; it's what he needed to do. He had gone over each and every job before and critiqued what he and other cops had done, what went right, and, more importantly, what went wrong. He made mental notes, it was a part of the plan–the plan to survive.

Things were happening so quickly Hogan's initial reaction was to avoid being hit by the gunfire. He did not have time to draw his own service weapon. According to Hogan's plan, he would first escape the confines of the car and get cover, and then he would draw and return fire. That is what he was trained to do. As he pulled on the door handle, he heard a shot.

POP.

This would be the last shot that Hogan recalled hearing. There was nothing but silence. Everything was quiet. Hogan heard something, or did he? There was a high-pitched noise, a ringing sensation. Hogan began to realize that the ringing was in his head, it was constant. The noise was coming from inside of him.

Oh my God, Oh my God. Oh my God. I'm hit. I'm hit.

Everything was black, the lights were out. The bright snow was gone. Hogan was temporarily blinded.

Unit 204: *Alright, coming from the center.*

Back-up was on the way.

Wow, it finally happened, I'm shot. I'm shot.

Hogan had expected that he would, sooner or later, be shot. He was not sure how he would react but would soon begin that journey. The darkness was vivid; it magnified Hogan's physical and emotional feelings. After realizing that he has been shot, Hogan felt a sudden sense of peace. Something was missing. The pain, there was no pain. Hogan began to identify his lack of feeling. The calm and silence, his sense of peace. Death. It finally happened! Hogan's first thoughts were of his father who had been murdered when he was just three years old.

Hello Dad, I'm gonna finally meet you.

His thoughts continued to focus on death. He thought of Tony Graffa, the Irvington police officer who was killed in the line of duty in 1984. He was a role model for Hogan, a cop who he admired. Graffa was someone who Hogan had looked up to as a young police officer, and his death had a profound effect on Hogan.

Hey Tony, here I come.

Hogan stopped. His thoughts radically changed direction. He thought about life. The life he and his fiancée were planning. They had been dating for seven years. Just the day before, they had celebrated the wedding shower; this was a part of a plan—a plan that did not include dying, at least not today.

Wait a minute, I gotta marry my girl.

Hogan's mind was racing. His thoughts went back and forth to his training, his plan for survival.

You know what this is. Get up Kenny, get up. Get mad.

Death's door, Ken Hogan was at death's door. Not for long.

Get mad Kenny, Get mad Kenny. Fight Back. Move, get up off
of the floor. Oh my God, this is bad. Get up off of the floor Ken.
Fight Kenny. FIGHT. Oh my God.

The feeling of peace that Hogan initially felt quickly gave way to a rushing sense of fear, which was a good thing. It meant that Hogan was still alive. He knew this and he kept talking to himself.

Move, get up. Get away.

Don't die Kenny . . . don't die . . . You can do this. You're not going to die today.

He played this over and over in his mind. The mantra was cathartic.

Don't die Kenny . . . don't die. Don't die Kenny.

Hogan sat up in the car. There still was no sensation of pain. His physical movement felt exaggerated, as if he were moving in slow motion. Suddenly he realized that two of his senses were not working. Ken Hogan could not see or hear anything. This realization was flooded with fear.

I can't see. My eyes, my eyes, I can't see.

Where is he? Where is the bad guy?

Fully aware that he was upright in the police car, Hogan could not determine where his assailant was. There was no time to feel pity or sorrow. Hogan had to do something. He had to survive. That was a part of his plan.

Hogan visualized in his mind's eye exactly where he was just before he was shot. What he was not aware of at that moment was how many times and where he had been shot and just how seriously he was injured. Four shots struck Hogan. The first hit him in the right hand, followed by two shots in his back–his right shoulder and spine at the base of his neck. The last shot was point-blank to the center of his head. The plan, the scenario, these were things that Hogan had gone over in his mind. He was prepared for this. This was part of his plan.

I know this. I know this. I know this. Do it, do it know.

As he tried to move, Hogan was acutely aware that his right arm was limp at his side. Still, he knew he must react. Trying to start the car, which he had turned off just before calling to the suspect, Hogan realized that he cannot use his right arm. Thankfully, the large key ring would help Hogan in starting the car without having to look for the guys; he never pulled the key from the ignition and dropped it to the floor, something he had done countless times before. Luck or consequence would play an integral part in Hogan's movements immediately after he was shot.

Fear was an emotion that Hogan had prepared for; he built fear into his scenarios. Fear was an ally at this point. It signaled that he was still alive and that he needed to act, he needed to act now.

I CAN'T SEE. Oh my God.

Call Kenny, call Headquarters. Do they know where I am?

Acting according to his plan, something that he consciously thought of, Hogan reached for the radio microphone with his left hand.

Just reach up and grab the mic, reach up Ken, its there. It's always there.

Hogan: *3 to quarters. I've been shot!*

Did they hear me? Is the radio working? Oh my God.

START THE CAR KENNY. START THE CAR. MOVE. MOVE. MOVE.

Using his left hand, Hogan started the car; the keys were in the ignition, and he hadn't dropped them to the floor like he normally would have, time and time again. Not this time. Not able to hear the engine, he could feel the vibration in the car as the engine came to life.

Drive Kenny, drive. Drive the dam car.

I CAN'T SEE. I CAN'T SEE. HOW AM I SUPPOSE TO DRIVE, I CAN'T SEE. JUST DRIVE, DRIVE.

Still not feeling pain, Hogan's thoughts returned to his assailant.

Where is he? Please don't let him get off any more rounds.

With the car running, Hogan sat upright and put the car into gear.

I have to get away. Don't let him shot again.

Hogan sensed the vehicle was moving and turned to his right. As a right-handed person, this was as much instinct as it was situational awareness based on his recollections of the scene prior to the shooting. As the police car began to move, Hogan realized that he was regaining his sight, he was able to see. The first image that he recalled was a police car heading in his direction.

Thank God, thank you God. Help. Help is here.

With the approach of the back-up unit, Hogan felt the light fading from his view. It was dark again, his sight was gone. He could feel the fear filling him again, but not pain. There was no pain. Hogan passed out. Again, the recognition, no pain. Hogan knew that if he did not fight, he would die.

OPEN YOUR EYES, KENNY. OPEN YOUR DAMN EYES.

In response to his subconscious, Hogan opened his eyes. Looking directly at his right hand, he saw that his hand was covered in blood.

My finger, my finger is gone. This guy shot off my finger.

The first shot to strike Hogan was to his right hand. The index finger was severed from the gunshot. Looking at his hand, Hogan began to sense that he was wet, everything was wet, his entire body. Fear crept back into the picture.

The sun, I can see the sun.

Not sure of what was happening; he pressed the accelerator in an attempt to keep the police car moving. Little did he know at that time, but the car had ridden up onto a snow bank and would not move. Ken was trying to escape; he was trying to complete his plan.

RUN, RUN, RUN KENNY.

HQ: Alright, units, he states he's been shot. Units assist officer.

HQ: Units assist officer.

HQ: 21st and Nelson.

Hogan: I've been shot!

HQ: We've got units coming down. 21st and Nelson. Officer's been shot there.

BEEP. BEEP. BEEP. (HQ sends out an alert tone from dispatch center)

HQ: Units, officer shot, Nelson and 21st.

Hogan pushed the alert button from inside his police vehicle located on the floor-mounted car radio–signaling the alert tone. This alert sends out an audible tone that signals distress. Every cop is familiar with the beep-beep-beep tone. Ken had practiced pressing this button numerous times, some with his eyes closed. This was all part of his plan, and he was putting it into action. He used his left hand to find the radio and continued to use his fingers, like a blind person relying on Braille; it worked, he pushed the button.

BEEP. BEEP. BEEP.

HQ: We're getting a signal from his vehicle now, units.

Don't let him get you. You gotta move. Get out of here. Don't let him get you again.

The other police car was an escape; it was a safe haven away from the shooter. Hogan decided that his best option at this point was to get to the other police car. Not realizing the extent of his injuries, Hogan planned on leaving his patrol car and running to the safety of the back-up unit. In his mind, Hogan ran to the other car. In his mind he ran.

> **HQ:** Black jacket and army pants from a witness over the phone.
> **HQ:** Through the yards 335 . . . 21st
> **204:** Out.

As information is relayed over the police radio, Hogan's thoughts were on getting out of the radio car and into the police car that he saw just moments before. In his mind, he ran. Despite his injuries, Hogan managed to open the door of Car #7. Hogan visualized running to the safety of the other car. Little did he know that he barely made it from the car.

In reality, he stumbled from his car and was helped into the backseat of the second police car by Officer Dennis Doherty (Unit 204).

> **204:** 4 to quarters. You want me to take him to the hospital?

Doherty's voice is strong, authoritative; he's a veteran cop, just like Hogan. Hogan responded to Doherty's last radio transmission.
I'm dying Dennis, I'm dying, just take me.

> **HQ:** Everything should be there Dennis, just stand by there. Everything should be there momentarily.

Hogan was drifting in and out.
Dennis take me. Take me. I'm gonna die here. Take me.
Doherty had seen and heard enough. Hogan was in bad shape, real bad shape. Sounding less confident than before, Officer Doherty made the decision to transport Hogan to the hospital.

> **204:** I'm starting down there.

As Officer Doherty began to make his way to University Hospital, the State Trauma Center in Newark, Hogan lay in the backseat of the radio car bleeding severely from multiple gunshot wounds. Shot in the head, neck, shoulder, and hand, he felt the life draining from his body. Hogan continued to fight unconsciousness. He knew that passing out would mean death. His plan was to talk to Doherty in an effort to remain conscious.

Don't pass out Kenny. If you pass out you'll die. Don't pass out.
Don't die, I don't want to die.

As Hogan continued to talk to Officer Doherty, he felt more at ease. Fear was slipping away, but so was his consciousness. The last thing Officer Ken Hogan recalled before passing out in the rear of Officer Doherty's car was the entrance way sign for the UMDNJ Emergency Room.

HQ: Units we just got a call—22nd Street between 20th and 19th Avenue.
Male shot himself in the head there.

As Officer Doherty pulled into the parking lot of the Emergency Room, he expected that a trauma team would be waiting for him. No one, there was no one waiting. No one was there to help. Hogan lay in the back seat of Doherty's radio car. He was bleeding out. Ken Hogan was dying.

Doherty sprang into action, determined to help. He ran into the Emergency Room and yelled

COP SHOT. COP SHOT.

The staff, the nurses, doctors, anyone nearby rushed to assist. Hogan was brought into the Emergency Room. He was at death's door, but not for long, not today. There was a bond between medical personnel and the cops and firefighters who came in and out of emergency rooms across the nation. UMDNJ was no different. Cops were like one of their own. They would help Ken, they would help him.

ONE MAN'S RESILIENCY

As the radio car driven by Officer Dennis Doherty rolled into the receiving area for College Hospital (a common identifier for the

University of Medicine and Dentistry), Hogan saw the sign "Emergency Entrance." His emotions were calm; "I was at peace, no pain, no fear. It was quiet, all quiet." Through the fog of the shooting and not fully aware of what was happening, he heard a voice. A women's voice. As he opened his eyes, Ken saw Roseanne, his fiancée, and mother at his side. Ken sensed Roseanne's strength; she kept any visible sign of emotion in check.

As Ken lay on the stretcher, he began to cry. The tears were in response to the pain, physical and emotional. A doctor introduced himself to Ken and his fiancée. "Hello Ken, I'm Doctor Hunt." The doctor was Charles David Hunt, and Hogan recalled Dr. Hunt as appearing "stern," but "he had all the confidence of a winning Yankee team." For the first time, Hogan was informed that he had been shot four times and that the bullets were likely still in his body.

In the Emergency Room, the doctor offered Hogan two options. Hogan's first thought was not pleasant. I only have two options? The options were to stabilize him and try to remove the bullets in surgery at a later time or to bring him into surgery at that moment. The choice for Hogan came instinctually: Now. He wanted the doctor to operate without waiting.

After surgery, Hogan's emotions were running rampant. Physically he could not see or feel. Fear, he was feeling fear. Recognizing the fear, his mind allowed him to reach back, back in time, back to Sister Maureen, back when he was a kid. Hogan hated fear. Fear was weakness, and he needed all the strength he could muster to live through this. Ken Hogan's past was the foundation for the resiliency that would help him through the minutes, hours, weeks, and months of recovery that lay ahead.

After the surgery, Dr. Hunt assessed Hogan. The doctor checked for movement in Hogan's legs and feet: "Can you feel this?" "Yes." Can you move your toes?" "Yes." "Do you have pain anywhere else?" "My head, my head is killing me." Hogan recalled that the pain was overwhelming, but that meant that he was alive, and alive was good. Alive was good. Ken truly believed that he could endure the pain, and as long as he felt pain, he was recovering.

The period of recovery for Hogan was long and arduous. His fiancée and his career were in the forefront of his mind. Would he get married? Would he still be a cop? The need for strong pain medication worried him; would he become an addict? The physical rehabilitation was not the first step in Hogan's recovery. Every cop in northern New Jersey knows the reputation of College Hospital, the State Trauma Center where he was treated. Hogan was in excellent hands, and the prognosis for his physical recovery was good. His mental state was of primary importance immediately following the surgery. In the first few days after the shooting, doctors and clinicians cautioned Ken to keep an open mind about his emotions, specifically about anger, his anger, with himself and about his future.

Hogan's initial recovery, both mentally and physically, would happen in a world wind of activity. The shooting occurred on January 24, 1994. He was promoted to sergeant on February 28. Forty-one days after being shot four times, Ken walked his fiancée Roseanne down the aisle. By September 1996, Ken Hogan returned to full duty and to the work that he was so passionate about.

Ken Hogan has been recognized time and time again for his compassion and courage over the course of a 25-year career. His experiences as a young boy growing up in a dysfunctional environment were not enough to keep him down. His optimism, and a strong belief in himself, offered hope despite of difficult times.

The social utility of a story such as Ken Hogan's can be found in his optimism. To those who know him, it is evident that his optimism is infectious. Ken has always had a strong belief in himself. Not arrogant or self-righteous, his keen sense of situational awareness as a police officer has helped to prepare him as he visits with police officers who have been shot or wounded. There is no panacea for resiliency. There is no formula or workout that offers resiliency. The Ken Hogan story is the story of just one man, one man, who in the face of adversity, pain, and doubt would not give up. Ken Hogan had a plan.

Chapter 7

RESILIENCY IN HIGH RISK GROUPS: A QUALITATIVE ANALYSIS OF LAW ENFORCEMENT AND ELITE MILITARY PERSONNEL

GEORGE S. EVERLY, JR. AND ANNE LINKS

DEFINING RESILIENCE: THE JOHNS HOPKINS PERSPECTIVE

Resiliency has most recently emerged as a critical phenomenon in areas such as child development, disaster mental health, law enforcement, and the military. In this chapter, we report on our initial attempt to provide qualitative insight into the statistical models we have already developed in our quest to reveal the underpinnings of human resiliency.

Resiliency may be thought of as the ability to rebound, or bounce back, from adversity (Everly, 2009; Everly, Strouse, Everly, 2010; Kaminsky, McCabe, Langlieb, & Everly, 2007). The aforementioned definition suffers from combined constructs, however. In order to provide greater clarity, we reformulated the construct of human resiliency into three distinct phenomena and proffered this reformulation as the Johns Hopkins Resistance, Resiliency, and Recovery Model (Kaminsky, McCabe, Langlieb, & Everly, 2007). This model represents a tripartite temporal continuum.

Resistance, the initial element, refers to the ability of an individual, a group, an organization, or even an entire population to literally resist manifestations of clinical distress, impairment, or dysfunction associated with critical incidents, terrorism, and even mass disasters. Resistance may be thought of as a form of psychological/behavioral *immunity* to distress and dysfunction.

Historically, this element was conspicuous in its absence. The notion of creating resistance represents a proactive step in disaster preparedness, emergency mental health, and training for high risk professional groups. Notions of "psychological immunization" and "psychological body armor" are engendered by the introduction of this intervention to the pre-incident phase of the temporal continuum.

Resilience, the second element, refers to the ability of an individual, a group, an organization, or even an entire population to rapidly and effectively rebound from psychological and/or behavioral perturbations associated with critical incidents, terrorism, and even mass disasters.

Collectively, we believe that resistance and resilience may be facilitated by two "active ingredients": *expectancy and experience*. The primary formulation that serves as the basis for this notion resides in the work of Albert Bandura (self-efficacy). Bandura's work (1997) is summarized in his magnum opus on self-efficacy and human agency. Bandura defines the perception of self-efficacy as the belief in one's ability to organize and execute the courses of action required to achieve necessary and desired goals. This perception of control, or influence, Bandura points out, is an essential aspect of life itself: "People guide their lives by their beliefs of personal efficacy" (Bandura, 1997, p. 3). He goes on to note: "People's beliefs in their efficacy have diverse effects. Such beliefs influence the courses of action people choose to pursue, how much effort they put forth in given endeavors, how long they will persevere in the face of obstacles and failures, their resilience to adversity, whether their thought patterns are self-hindering or self-aiding, how much stress and depression they experience in coping with taxing environmental demands, and the level of accomplishments they realize" (Bandura, 1997, p. 3).

Bandura (1977, 1982, 1997) has described four sources that affect the perception of self-efficacy and are particularly relevant in terms of the resistance/resilience model. They are: (1) *performance*—"Enactive attainments provide the most influential source of efficacy information. Successes raise efficacy appraisals, repeated failures lower them" (Bandura, 1982, pp. 26–27); (2) *vicarious experience*—"Self-efficacy appraisals are also partly influenced by vicarious experiences. Seeing or visualizing similar others perform successfully can raise self-percepts of efficacy in observers that they too possess the capabilities to master comparable activities. . . . By the same token, observing others to be of similar competence fail despite high efforts lowers observers judgments of their own capabilities and undermines their efforts" (Bandura, 1982, p. 27); (3) *verbal persuasion and support*–Verbal persuasion comprises such things such as suggestion, coaching and education; and (4) *physiological/affective arousal*–"People rely partly on their state of physiological arousal in judging their capabilities and vulnerability to stress. Because unusually high arousal usually debilitates performance, individuals are more likely to expect success when they are not beset by aversive arousal. Fear reactions generate further fear through anticipatory self-arousal. . . . People can rouse themselves to elevated levels of distress that produce the very dysfunctions they fear. Treatments that eliminate emotional arousal heighten perceived efficacy with corresponding improvements in performance" (Bandura, 1982, p. 28).

Recovery, the third and final element in the Hopkins model, refers to treatment and rehabilitation. It completes the tripartite and comprehensive continuum but is not relevant beyond enumeration to this chapter.

STRUCTURAL MODELING

Research by our group has sought to elucidate the nature of human resilience from a structural modeling perspective as a means of discovering the infrastructure of human resistance and resiliency. The primary tools employed were correlational and structural modeling analyses. Structural modeling is a technique for testing and estimating causal relations using a combination of statistical data and qualitative assumptions.

Smith, Everly, and Johns (1993), Smith, Davy, and Everly, (1995, 2006, 2007), Everly, Smith, and Welzant (2008), and Everly, Smith, and Lating (2009) used statistical modeling to better understand the mechanisms of action that actually serve to account for the ability to resist distress or rebound after adversity.

Smith, Davy, and Everly (1995, 2006, 2007) discovered that stressors (i.e., challenging or potentially pathogenic factors at work) exerted little direct effect on factors such as job performance, job satisfaction, and intentionality to seek alternative employment. Rather, it was revealed that cognitive and affective states seemed to serve as mediators either buffering or increasing vulnerability. This finding was consistent with earlier research by Smith, Everly, and Johns (1993), which found that cognitive affective states (attitudes) served as mediators in the stressor to physical illness process.

Everly, Smith, and Welzant (2008) researched the variable of positive emotionality as a contributor to resistance and resilience. Positive emotionality may be thought of as the presence of ambient positive emotions, as well as the ability to express positive emotions in the wake of adversity. That investigation assessed the relationship between the expression of ambient emotions (within the last few weeks) on measures of burnout, job satisfaction, perceived performance, and intention to leave one's job in a study consisting of a randomized sample of 2,500 out of approximately 91,333 potential subjects. Four hundred eighty-nine and 491 usable responses were received for inclusion in the analyses. Correlational analyses revealed that expressed positive emotions, as well as expressed negative emotions, were related to the outcome variables in a significant but complementary manner so as to predict burnout and job-related outcome. More specifically, positive emotions appear to support job satisfaction and performance, whereas negative emotions appear to be predictors of burnout and intentions to leave the job.

Everly, Smith, and Lating (2009) utilized the responses of the 491 individuals employed in public accounting to further investigate cognitive-affective dynamics in human resistance and resiliency. Results indicate that the cognitive-affective domain is an essential determinant of burnout, job dissatisfaction, turnover intention, and performance. Furthermore, cognitive states appear to exert their effect through affective arousal that subsequently appears to have a defining role in the development of the aforementioned variables.

The aforementioned investigations demonstrate the importance of attitudinal variables in resistance and resilience. In an effort to contextualize these findings, we extended the investigations in an attempt to identify specific cognitive-affective states of relevance.

METHOD

One hundred and fifty-eight individuals from three high risk professional groups volunteered for the present study (20 former or current Navy SEALs, 25 current SWAT team members, and 113 federal law enforcement agents). Participation was in combination with interactional educational processes and focus groups. The average age of participants was 42 years (SD 10.8), and the average years of active duty experience was 14.5 years (SD 5.8).

Participants were asked in open end question format to answer the following five questions:

1. What is the key to being immune to stress?
2. What is the key to bouncing back from excessive stress?
3. What is your greatest strength (key to success)?
4. If most people have a weakness that makes them vulnerable to excessive stress, what is it?
5. What is the key to motivating people to help them be successful?

Two individuals read through the accumulated data and categorized it into qualitatively homogenous categories for ease of reporting.

RESULTS

The categorized data are reported as percentages and are as follows (percentages exceed 100 percent because of multiple answers):

1. What is the key to being immune to stress?

- Positive attitude: 30%
- Training: 28%
- Healthy lifestyle: 28%
- Having an outlet, or hobby: 24%
- Support network, including leadership: 8%
- Other: 10%

2. What is the key to bouncing back from excessive stress?

- Positive attitude: 33%
- Having an outlet, or hobby: 26%
- Support network, including leadership: 25%
- Healthy lifestyle: 8%
- Removal from stressful situation: 18%
- Other: 16%

3. What is your greatest strength *(key to success)*?

- Work ethic: 30%
- Tenacity: 29%
- Positive attitude: 22%
- Strong values: 21%
- Adaptability: 9%
- Other 13%

4. If most people have a weakness that makes them vulnerable to excessive stress, what is it?

- Lack of perspective: 41% (lack of tenacity, lack of preparation)
- Negative attitude: 24%
- No outlet: 12%
- No preparation: 9%
- Isolation: 5%
- Other: 18%

5. What is the key to motivating people to help them be successful?

- Encouragement: 44%
- Lead by example: 22%
- Training and experience: 18%
- Communication: 11%
- Other: 23%

There did not appear to be significant variation in responses among the three professional groups, other than the SEALs seemed to more consistently endorse positive attitude and training as the key variables undergirding resistance and resilience.

DISCUSSION

A review of the present data reveals positive attitudes are perceived as being the most important attributes that determine both resistance and resilience as we have previously defined them. These findings are confirmatory of the structural modeling analyses previously described, that is, attitude plays a critical causal role in the determination of success and growth on the one hand versus distress and failure on the other hand.

Optimism, tenacity, and interpersonal support seem to be essential elements in our quest to understand human resistance and resiliency. Fortunately, all three elements can be learned and enhanced through training, we believe. For example, critical incident response teams have become a virtual industry standard in law enforcement. Their primary purpose is to mitigate exposure to adversity through formalized mechanisms of interpersonal support (Everly & Mitchell, 1999), and the effectiveness of critical incident stress management protocols has been demonstrated (Boscarino & Adams, 2008). Clearly optimism can be learned (Seligman, 1991).

It was revealing that in the verbal narratives obtained in the focus groups with Navy SEALs, albeit limited, a subtle yet powerful difference in the definition of "optimism" was revealed. Optimism is often perceived of as a hopefulness, an expectation of something positive, whereas among SEALs, optimism was viewed as a mandate for achievement.

At the "systems' level," it is believed that the most effective way to create a "culture of resistance and resilience" is to train first-line managers and supervisors in "resilient leadership" (Everly, Strouse, & Everly, 2010). This concept is consistent with the notion of the tipping point as espoused by Malcom Gladwell (2000). Resilient leadership refers to those leadership characteristics that encourage others to be resistant and resilient in the face of adversity. Resilient leadership depends largely on four critical attributes: (1) integrity, (2) open communications, (3) decisiveness, and (4) optimism. Again, we believe that all of these qualities can be taught both effectively and efficiently. Similarly, we know that poor leadership can be a significant source of organizational and operational distress (Hoge et al., 2004).

Appreciation is extended to Captain Donald Hinsvark (USNR, Retired), SA Joshua Knapp, and Captain Michael Cobb without whom the investigation would not have been possible.

REFERENCES

Bandura, A. (1977). Self-efficacy: Toward a unifying theory of behavior change. *Psychological Review, 84,* 191–215.

Bandura, A. (1982). The self and mechanisms of agency. In J. Suls (Ed.), *Psychological perspectives on the self* (pp. 3–39). Hillsdale, NJ: Erlbaum.

Bandura, A. (1997). Self-efficacy: *The exercise of control.* New York: Freeman.

Boscarino, J. A., & Adams, R. E. (2008). Overview of findings from the World Trade Center disaster outcome study: Recommendations for future research after exposure to psychological trauma. *International Journal of Emergency Mental Health, 10*(4), 275–290.

Everly, G. S., Jr. (2009). *The resilient child.* New York: DiaMedica.

Everly, G. S., Jr., & Mitchell, J.T . (1999). *Critical incident stress management.* Ellicott City, MD: Chevron.

Everly, G. S., Jr, Smith, K., & Lating, J. (2009). Rationale for cognitively based resilience and psychological first aid (PFA) training: A structural modeling analysis. *International Journal of Emergency Mental Health, 11*(4), 249–262.

Everly, G. S., Jr., Smith, K. J., & Welzant, V. (2008). Cognitive-affective resilience indicia as predictors of burnout and job-related outcome. *International Journal of Emergency Mental Health, 10*(3), 185–190.

Everly, G. S., Jr., Strouse, D. A., & Everly, G. S., III (2010). *Resilient leadership.* New York: DiaMedica.

Gladwell, M. (2000). *Tipping point.* New York: Little, Brown.

Hoge, C. D., Castro, C. A., Messer, S., McGurk, D,. Cotting, D., & Koffman, R. L. (2004). Combat duty in Iraq and Afghanistan, mental health problems, and barriers to care. *New England Journal of Medicine, 351,* 13–22.

Kaminsky, M. J., McCabe, O. L., Lans:lieb, A., & Everly, G. S. Jr. (2007). An evidence-informed model of human resistance, resilience, and recovery. The John Hopkins outcomes-driven paradigm for disaster mental health services. *Brief Therapy & Crisis Intervention, 7,* 1–11.

Seligman, M. E. P. (1991). *Learned optimism.* New York: Knopf.

Smith, K. J., Davy, J. A., & Everly, G. S., Jr. (1995). An examination of the antecedents of job dissatisfaction and turnover intentions among CPAs in public accounting. *Accounting Enquiries, 5*(1), 99–142.

Smith, K. J., Davy, J. A., & Everly, G. S. (2006). Stress arousal and burnout: A construct distinctiveness evaluation. *Psychological Reports, 99,* 396–406.

Smith, K. J., Davy, J. A., & Everly, G. S., Jr. (2007). An assessment of the contribution of stress arousal to the beyond the role stress model. *Advances in Accounting Behavioral Research, 10,* 127–158.

Smith, K. J., Everly, G. S., & Johns, T. R. (1993). The role of stress arousal in the dynamics of the stressor to illness process among accountants. *Contemporary Accounting Research, 9*(2), 432–449.

Chapter 8

LEADING FOR RESILIENCE IN HIGH RISK OCCUPATIONS

Paul T. Bartone and Charles L. Barry

INTRODUCTION

Effective and reliable worker performance is important in every job, but in some jobs the risk of failure is greater than in others. High risk occupations are dangerous ones that routinely involve life or death consequences. Workers in high-risk occupations, should their performance slip, can bring about death, injury, or other serious negative consequences on themselves, fellow workers, clients, outside communities, or the environment. By this definition, a wide range of occupations would qualify as high risk, from police and security to emergency responders and rescue personnel, a range of heavy industrial and mining occupations, transportation and traffic control functions, many medical jobs, and many jobs in energy. Therefore, it is critical that every effort be made to understand the causes of human breakdown in high risk occupations and address these with programs to mitigate the risk of failure. To that end, the present chapter examines the military as a case study of a high risk occupation. With this as our reference point, we draw general lessons regarding the key sources of stress in high risk occupations, how they impact on individual workers and the organization as a whole, and what can be done to counter these stressors and reduce the risk of performance decrements and failures.

The military occupation is certainly a stressful one and one that carries high risks, whether in times of peace or war. In wartime, highly lethal weapons can do great harm, and friendly and innocent parties can easily end up as victims. In peacetime, too, dangers from the handling of heavy equipment and weapons systems, as well as responding to a whole range of human and natural disasters, also place military personnel in harm's way. In recent years, military personnel are deploying more often and for longer periods of time on missions that are multidimensional, ambiguous, and often changing (Alford & Cuomo, 2009). Such risk factors and stressors are found in many occupations outside the military as well (see Violanti, Chapter 10, this volume). These on-the-job stressors can lead to a range of health problems and performance decrements in workers. But not everyone reacts in negative ways. The majority remains healthy and continue to perform well even in the face of high stress levels. We need a better understanding of what factors contribute to such stress resilience.

This chapter focuses attention on mental hardiness, an important pathway to resilience (Kobasa, 1979; Maddi & Kobasa, 1984). Research over the past 25 years has confirmed that psychological hardiness is a key stress-resiliency factor (Bonanno, 2004; Watson, Ritchie, Demer, Bartone, & Pfefferbaum, 2006). People high in psychological hardiness show greater commitment–the abiding sense that life is meaningful and worth living; control–the belief that one chooses and influences his or her own future; and acceptance of challenge–a perspective on change in life as something that is interesting and valuable. We begin with an essential first step: clarifying the major stress factors that are salient in modern military operations. Next, we give a brief summary of the theory and research behind the hardiness construct. Finally, we provide a number of suggestions for how to increase hardiness and stress resilience in organizations, primarily through leader actions and policies. By setting the conditions that increase mental hardiness, leaders at all levels can enhance human health and performance while preventing many stress-related problems before they occur.

PSYCHOLOGICAL STRESS FACTORS
IN MODERN MILITARY OPERATIONS

The military occupation exposes its workers to a wide range of stressors. Combat-related stressors are perhaps the most obvious and extreme ones, and these tend to get the most attention (e.g., Hoge et al., 2004; Marlowe, 1986;). But combat is not the only source of stress on military operations, and may not be the most potentially damaging. Modern military operations entail a wide range of challenges and potential stress factors (Krueger, 2008). In the post-Cold War era, the number of peacekeeping, peacemaking, humanitarian, and other kinds of operations have increased dramatically, whereas military force levels have shrunk. Partly as a result of substantial 1990s force reductions, deployments are more frequent and longer in duration than in times past, especially for U.S. Army personnel. This in turn has brought other changes in military units, including more training exercises, planning sessions, and equipment inspections in preparation for deployment. All of this adds to the workload and pace of operations on the home front (Castro & Adler, 1999). More intense work schedules and frequent deployments also force more family separations, a well-documented stressor for service members (Bell, Bartone, Bartone, Schumm, & Gade, 1997; op den Buijs, Andres, & Bartone, 2010).

One place to look for reducing the stress associated with high risk operations is to lessen the workload while increasing rest periods and recovery time. In the military, this means largely reducing the frequency and duration of deployments. Unfortunately, strategic imperatives and troop shortages may prevent this. The military is not alone in this regard; the same is true (at least at times) in other occupations and contexts. For example, following the 9/11 terrorist strike on the World Trade Center, fire, police, and other emergency personnel maintained continuous operations around the clock with the goal of locating possible survivors, as well as restoring essential services to the affected areas. Thousands of disaster response workers were involved in rescuing victims and restoring basic services in New Orleans following Hurricane Katrina in August 2005. In such crisis situations, continuous operations and extreme efforts are necessary to save lives; easing the pace of work may be considered unacceptable or even unethical. However, when operations become long term, as with many military operations, workload requirements should be realigned with what the existing workforce can reasonably sustain. Leaders and managers need to be aware of the limits of human endurance and the importance of adequate rest cycles.

In order to mitigate or counter the stressors associated with high-risk occupations like the military, it is important to begin with an analysis and clarification of the nature of the stressors encountered by workers on the job. In the case of military operations, extensive field research with deployed U.S. military units led to the identification of five primary psychological stress dimensions associated with modern military operations: (1) Isolation, (2) Ambiguity, (3) Powerlessness, (4) Boredom, and (5) Danger (Bartone, Adler & Vaitkus, 1998). Today, with the greatly increased frequency and pace of deployments for U.S. forces and the long work periods involved, a final important stress factor is workload or "operations tempo." These six dimensions are elaborated below.

1. *Isolation.* When military personnel deploy on operations, it is typically to remote areas, far from home and families. Reliable methods for communicating with home are often lacking. Most of the usual stress-relieving activities, such as exercise, athletics, sports, television, movies, games, and similar options, are not available. Email is usually not available, and traditional mail may be sporadic and can take weeks to deliver. Soldiers find themselves in strange lands and cultures, far from familiar surroundings. Although fellow soldiers are mostly known and familiar due to the U.S. Army's current reliance on unit rotation policies, there are still some individual replacements due to casualties and other unexpected depletions of essential unit strength. For individual replacements, the initial stress of social isolation can be more acute as they attempt to fit into an established group of friends. Also in many cases, deploying units are configured as task forces tailored for specific missions,which means many unit members are strangers who haven't worked together previously. Security and operational concerns (e.g., "force protection") often generate movement restrictions, as for example when troops are restricted from leaving their base camp. Troops may also be banned from interacting with the local populace and prevented from participating in familiar activities such as jogging for exercise or displaying the U.S. flag. Frequently, there are also multiple constraints on dress and activities. They have

few choices in their daily existence. Movement and communication restrictions also deter troops from learning about local culture and language and about resources that might be available locally. All of these factors contribute to a sense of social isolation.

2. *Ambiguity.* In modern military operations, a unit's mission, rules of engagement and situation are often unclear to the individual service member and can require rapid role changes from active combatant to neutral peacekeeper to humanitarian assistance provider. In the late 1990s, Marine Corps Commandant Charles Krulak described this mix as a "three-block war" in reference to the need for Marines (and soldiers) to be able to conduct full-scale military action, peacekeeping operations, and humanitarian relief within the space of three contiguous city blocks, sometimes engaging in two or all three on the same day (Krulak, 1999). It can be hard for soldiers to quickly shift to different rules of engagement (ROE) (e.g., Do I knock on the door or kick it down?); mental judgment cannot be reprogrammed instantly or be fully divorced from emotions. The role and purpose of military personnel can be unclear in these conditions. Also, the command structure is not always clear, a situation that readily arises for example when support units are realigned to different combat units due to changing operational conditions. Another factor adding to ambiguity is insufficient knowledge of host nation language and cultural practices, although predeployment training may provide some basic information in this area. There is also frequently a lack of knowledge regarding the military contingents of allies, as well as the status and authority of contractors (in particular, paramilitary security forces or private security contractors [PSCs]) in a multinational coalition force. In counterinsurgency operations, troops often face continuing uncertainty regarding who is an enemy and who is an innocent civilian. All of this contributes to a highly ambiguous environment.

3. *Powerlessness.* Related to ambiguity is the sense of powerlessness to bring clarity to or exercise control over one's own destiny on a day-to-day basis. Troops may wonder, for example, "When will we be back from patrol?" "When will we move out again?" "Will I be able to meet my wife on leave when I was told (and told her)?" "Will our unit go home when we were told we would?" Soldiers and small unit leaders are often equally powerless to alter their mission conditions. In hierarchical military organizations, there is always some sense of power-

lessness that is magnified the further down the institutional hierarchy one goes. Limitations on soldiers' movements and activities, already noted as a stress factor in terms of accentuating isolation, may not be readily understandable if the local situation appears benign and soldiers from other coalition countries or even U.S. civilians are not so restricted. Soldiers have little power to change the rules under which they are governed. Another contributing element can be the soldier's sense of being helpless to assist or improve the lot of the local population. Soldiers may see local people in need of help–wounded, ill, hungry–but be unable to provide needed assistance due to restrictive ROE, lack of supplies, or operational or political considerations. For example, operational requirements to keep the unit moving can interfere with establishing relationships of trust with the local population and community leaders. Returning time and again to the same location to reestablish security or perform some other mission such as interdicting or destroying illicit drugs further adds to the sense of being unable to solve the problem or complete the mission. All of this contributes to a potentially damaging sense of powerlessness that one has little control over the surrounding environment. Other studies have also identified powerlessness as a damaging influence for soldiers on peacekeeping operations. For example, Weisaeth and Sund (1982) found that in Norwegian soldiers serving in Lebanon under the United Nations Interim Force in Lebanon (UNIFIL) peacekeeping mission, the feeling of being powerless to act or intervene when witnessing some atrocity was a main contributor to posttraumatic stress disorder (PTSD) symptoms.

4. *Boredom.* Modern military missions frequently involve long periods of "staying in place" without much real work to do. A related situation is long periods of strenuous patrolling in areas where there is no enemy activity and no civilian population. As the weeks and months crawl by, a feeling of boredom grows. At a superficial level, boredom can be countered with more entertainment and sports activities. But the real problem of boredom is not about a lack of activities to engage in but rather a lack of *meaningful* or *constructive* work activities in which to engage. On many jobs and deployments, daily tasks can take on a repetitive dullness, with a sense that nothing important is being accomplished. This can be especially disturbing when the daily tasks are arduous, sustained, and unpleasant–such as patrolling in rugged ter-

rain and bad weather with heavy combat loads, and without any evidence of real threat or enemy presence. Troops can easily degenerate from a high state of alertness to rote marching, daydreaming of home or focusing on their current discomforts. When enemy actions or emergency situations then do arise, such troops are often caught by surprise and experience higher casualties. Yet keeping troops alert for days, weeks, or even months without enemy contact is a very difficult leadership challenge (Hancock & Krueger, 2010).

5. *Danger.* This dimension encompasses the real physical dangers and threats that are often present in the deployed environment, threats that can result in injury or death. Things like bullets, mines, bombs, or other hazards in the deployed setting are included here, as well as the risk of accidents, disease, and exposure to toxic substances. One of the most troublesome threats has long been what is known as indirect fire from aircraft, artillery, or mortars. The inability to see the enemy and therefore be able to take some action to stop the attack adds to the problem. Often troops are not even sure whether moving away merely makes them a more likely target. In current U.S. and coalition operations in Iraq and Afghanistan, similar hidden dangers take the form of suicide bombers, snipers, and "improvised explosive devices" (IEDs). The danger can be direct, posing a threat to the individual soldier, or indirect, representing a threat to his or her comrades. Exposure to or attending to severely injured or dead people, and the psychological distress this can bring, also adds to the sense of danger for troops.

6. *Workload.* At the macrolevel, this factor represents the increasing frequency and duration of deployments that many military units are experiencing. As units redeploy home, they are already scheduled for their next deployment, even one or two years into the future. At a more immediate level, most deployments are characterized by a "24-hour, 7-days-a-week" work schedule in which soldiers are always on duty, with little or no time off. Work-related sleep deprivation is common in the deployed environment. Training and preparation activities in the period leading up to a deployment also usually entail a heavy workload and extremely long days. The same is generally true for military units returning home from a deployment, when soldiers must work overtime to ensure that all vehicles and equipment are properly cleaned, maintained, and accounted for. Often this means that antici-

pated family time gets truncated or sacrificed altogether.

Understanding these key sources of stress makes it possible to develop more focused and effective mitigation approaches. While training at the individual level to increase stress resiliency is possible and can certainly help, it is important to remember that many other factors at various levels also influence resilient responding to work-related stress. Figure 8.1 lists some of these factors at the individual, organizational policy and organizational structure levels. Although the individual level is important, it is not the only one. Wherever possible, stressors at the organizational policies, procedures, and structural levels should be addressed and mitigated as part of the overall corporate stress management strategy. Too often organizations fall back on the easier tactic of focusing exclusively on individuals, for example offering more "stress management" or prevention training.

Individual factors that are relevant for selection and training include social background factors, personality (including psychopathology), previous experience and education, maturity, intelligence, physical fitness, and family circumstances. Organizational policies also can exert an important influence on resilience or how the organization and its members respond to challenging or stressful events. Here, it is useful to distinguish between "macrolevel" policies, such as agency rules, regulations and directives, mission statements, deployment and rota-

Factors that can influence stess resilience

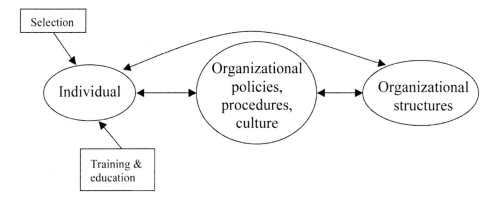

Figure 8.1. Many factors at multiple levels can influence resilient, healthy responding to stress.

tion policies, ROE and so on, and "microlevel" policies such as small unit policies, leader directives and communications, training schedules and programs, and the like that are worked out locally. At the same time it is important to recognize that some (but not all) microlevel policies and procedures are influenced rather directly by larger, macrolevel policies and standards. For example, if the Secretary of Defense or Chief of Staff of the Army should decide as a matter of policy that troop deployments into a war zone will last for 18 months or more, there is little room for small unit leaders to influence that policy.

Organizational structural factors also have an influence on how the organization responds to challenges. The size, type, and configuration of units may be more or less appropriate for the demands of a given environment at a particular time. Other structural considerations include such matters as where units are based geographically, how they are staffed or manned, the ratio of leaders to troops, and the integration of National Guard and Reserve forces, as well as joint and coalition forces. The integration issue here applies both in the context of specific missions, as well as regarding extended alliances (e.g., NATO ISAF–International Security Assistance Force–in Afghanistan). The arrows in Figure 8.1 serve as a reminder that these different major factors interact and influence each other as well. For example, organizational policies clearly influence (and in some cases determine) structures, while existing structures, force levels, and types have an influence on policies that are developed and implemented regarding their utilization. Structures and policies have an influence on individuals in myriad ways, as for example when force structures and rotation policies determine when and for how long an individual will be deployed. The horizontal bar labeled "Resources" at the bottom of Figure 8.1 is meant to indicate that all of these factors–individual, organizational policies, and organizational structures–are critically influenced by resource considerations. Budgets are limited, and what is done in any one area hinges to some degree on available time and money.

What tools, strategies, or coping mechanisms can be applied in order to increase resiliency or resistance to these stressors, both at the individual and unit levels? We focus below on the psychological style known as mental hardiness, and we discuss how leaders can increase individual and group resilience under stress by drawing lessons from hardiness theory and research.

Mental Hardiness

The "hardiness" construct first described by Kobasa in 1979 provides valuable insight for understanding highly resilient stress response patterns in individuals and groups. Hardiness was originally described as a personality trait or style that distinguishes people who remain healthy under stress from those who develop symptoms and health problems (Kobasa, 1979; Maddi & Kobasa, 1984). Hardy persons have a high sense of life and work *commitment,* have a greater expectation of *control,* and are more open to change and *challenges* in life. They tend to interpret stressful and painful experiences as a normal aspect of existence, a part of life that is overall interesting and worthwhile. While early tools for measuring hardiness had a number of problems (Funk, 1992), there are now excellent, reliable, and valid instruments for assessing the hardiness construct and the facets of commitment, control, and challenge (Bartone, 1995, 2007; Bartone, Roland, Picano, & Williams, 2008; Hystad, Eid, Johnsen, Laberg, & Bartone, 2010).

Although mental hardiness is relatively stable over time and across situations, there is good evidence that hardiness levels can be increased as a result of experiences and training (Maddi, Harvey, Khoshaba, Fazel, & Resurreccion, 2009; Zach, Raviv, & Inbar, 2007). So, although mental hardiness is relatively stable, it is not an immutable trait. Rather, hardiness is a generalized style of functioning that continues to be shaped by experience and social context throughout the lifespan. It includes cognitive, emotional, and behavioral features, and it characterizes people who stay healthy under stress in contrast to those who develop stress-related problems. The hardy-style person is courageous in the face of new experiences as well as disappointments and also tends to have a strong sense of self-efficacy or personal competence. Some studies suggest that hardiness, together with certain life experiences, operates to increase the sense of self-efficacy (Azar, Vasudeva, & Abdollahi, 2006). The high-hardy person, although not immune to the ill effects of stress, is thus robust and resilient in responding to highly stressful conditions.

The concept of hardiness is theoretically grounded in the work of existential philosophers and psychologists including Heidegger (1986), Frankl (1960), and Binswanger (1963). It involves the creation of meaning in life, even life that is sometimes painful or absurd, and having the courage to live life fully despite its inherent pain and futility. It is a generalized cognitive framework that affects how one views the self, others, work, and even the physical world. In Heidegger's existential terms, this is the *Eigenwelt,* the self or "I" world, the *Mitwelt,* the "with" or social world, and *Umwelt,* the "around" or physical world. In one of his early articles, Maddi (1967) outlined a progenitor of the hardy personality type and contrasted it with the nonhardy "existential neurotic" (Maddi, 1967). He used the term "ideal identity" to describe the person who lives a vigorous and proactive life, with an abiding sense of meaning and purpose, and a belief in his own ability to influence things.

Since the appearance of Kobasa's 1979 article, an extensive body of research has accumulated in support of the hypothesis that hardiness protects people in various occupational groups and high stress jobs against the ill effects of stress on health and performance. Studies have generally confirmed that hardiness operates as a significant moderator or buffer of stress (Contrada, 1989; Kobasa, Maddi & Kahn, 1982; Roth, Wiebe, Fillingim, & Shay, 1989; Wiebe, 1991). Hardiness has also been identified as a moderator of combat exposure stress in Gulf War soldiers (Bartone, 1993, 1999, 2000), U.S. Army casualty assistance workers (Bartone, Ursano, Wright, & Ingraham, 1989), U.S. peacekeeping soldiers (Bartone, 1996; Britt, Adler, & Bartone, 2001), Israeli officer candidates (Westman, 1990), Israeli soldiers in combat training (Florian, Mikulincer & Taubman, 1995), and Norwegian navy officer cadets (Bartone, Johnsen, Eid, Brun, & Laberg, 2002). For example, soldiers who develop PTSD symptoms following exposure to combat stressors are significantly lower in hardiness compared with those who don't get PTSD. Under low-stress conditions, troops high in hardiness report about the same level of PTSD symptoms as those low in hardiness. However, under high stress conditions, those high in hardiness report significantly fewer PTSD symptoms than those low in mental hardiness. These results provide additional evidence that those who are high in the qualities of hardiness are more resistant to the ill effects of extreme stress.

Leader Influence on Mental Hardiness

How does hardiness increase resiliency to stress? While the underlying mechanisms are still not fully understood, a key aspect of the hardiness resiliency process involves the interpretation or the meaning that individuals attach to events around them and to their own place in the world of experience. High-hardy people typically interpret life experience as (1) overall interesting and worthwhile, (2) something they can exert control over, and (3) challenging–presenting opportunities to learn and grow.

The power of psychological hardiness to mitigate stressful experiences is related most of all to the positive interpretations or framings of such experiences that are typically made by the hardy person. If a stressful or painful experience can be cognitively framed and made sense of within a broader perspective, which holds that all of existence is essentially interesting, worthwhile, a matter of personal choice, and providing valuable opportunities to learn and grow, then the stressful experience can have beneficial psychological effects instead of harmful ones. For example, in a study of U.S. Army soldiers deployed to Bosnia, those high in hardiness were more likely to perceive their work as meaningful (Britt, Adler, & Bartone, 2001). Additionally, the high-hardy person is more accepting or "forgiving" of a certain amount of disruption or pain as part of existence and prefers to look to the future rather than dwell on the past.

In organized work groups such as the military, this process of "meaning-making" can be more readily influenced by leader actions and policies. Military units by their nature are group-oriented and highly interdependent. Common tasks and missions are team actions, and the hierarchical authority structure frequently puts leaders in a position to exercise substantial control and influence over subordinates. By the policies and priorities they establish, the directives they provide, the advice and counsel they offer, the stories they tell, the amount of accurate and timely information they disseminate, and, perhaps most important, the examples they set, leaders can alter the manner in which their subordinates interpret and make sense of experiences. In these ways, leaders may, for example, better prepare subordinates for their first combat engagement. Such leader policies and actions may also better protect against the buildup of postengagement stress or PTSD.

Many authors have commented on how social processes can influence the creation of meaning by individuals in positive or negative directions. For example, Irving Janis (1972) used the term "group-

think" to describe how people in groups can come to premature clo-
sure on issues, with multiple individuals conforming to whatever is the
dominant viewpoint in the group. Similarly, Gordon Allport (1985),
noted American personality psychologist, viewed individual meaning
as often the result of social influence processes. Sociologists Berger and
Luckmann (1966) take an even more comprehensive view of the influence of social factors and "constructions" on the creation of individual
meaning, arguing that "reality" or perceptions of individuals reflect the
incorporation into the individual mind of social definitions of the
world. In turn, the organizational psychologist Karl Weick (1995)
describes some of the processes by which organizational policies and
programs can influence how individuals within the organization "make
sense" of or interpret their experiences, particularly at work. For
example, meetings and discussions at work provide key opportunities
for shared "framing" of experience and sensemaking. Peers, leaders,
and entire work units or organizational cultures can influence how
experiences get interpreted. But leaders are particularly influential in
shaping individual views.

The Israeli psychologist R. Gal (1987) is another who has argued
for the influence of leaders on subordinates, primarily along psychological pathways. He posits that a key influence mechanism involves
leaders increasing the sense of commitment in subordinates, including
commitment to the group, the mission, and the larger goals of the
organization. While not using the term "resilience," Gal believes this
leader-commitment-enhancement process to be especially important
for maintaining worker morale and performance during especially
demanding, high risk operations.

From our own perspective, leaders who are themselves high in hardiness and self-aware enough to understand the value of the kinds of
frames they use for making sense of experience can encourage those
around them to process stressful experiences in ways characteristic of
high-hardy persons. In a small-group context, leaders are in a unique
position to shape how stressful experiences are understood by members of the group. For example, leaders who are high in hardiness likely have a greater impact in their groups under high stress conditions,
when by their example, as well as by the explanations they give to the
group, they encourage others to interpret stressful events as interesting
challenges that can be met and in any case as opportunities to learn

valuable lessons for the future. This process, as well as the positive result (a shared understanding of the stressful event as something worthwhile and to learn from), could be expected to generate an increased sense of shared values, mutual respect, and cohesion. Further support for this interpretation comes from a study showing that hardiness and leadership interact to affect small-group cohesion levels following a rigorous military training exercise (Bartone, Johnsen, Eid, Brun, & Laberg, 2002). This interaction effect signifies that the positive influence of leaders on the growth of unit cohesion is greater when hardiness levels in the unit are high. This suggests that effective leaders increase group solidarity or cohesion at least in part by encouraging positive shared interpretations of stressful events when they occur.

SUMMARY AND RECOMMENDATIONS

The military is a high risk, high stress occupation. As with other high risk occupations, in order to reduce stress-related performance and health problems, it is important on the organizational side to work preventively to build up resilience and stress resistance of individuals and groups. This chapter identifies the key underlying stress factors in complex military operations—stress factors such as isolation, powerlessness, and heavy workload that also characterize other high risk occupations.

Based on sound theory and scientific support, we apply the concept of psychological hardiness to show how leaders can foster positive, resilient responding throughout their organizations. Efforts to increase resilience should span leader actions as well as organizational policies and programs at all levels.

Recommendations for Leaders

In high risk occupations such as the military, where individuals are regularly exposed to a range of stressors and hazards, leaders are in a unique position to shape how stressful experiences are processed, interpreted, and understood by members of the group. The leader, who by example, discussion, and established policies communicates a positive interpretation of shared stressful experiences, exerts a positive influence on the entire group in the direction of his or her interpretation of experience-toward more resilient and hardy sense making. Leaders can increase mental hardiness and resilient responding in several ways:

- *Set a clear example,* providing subordinates with a role model of the hardy approach to life, work, and reactions to stressful experiences. Through actions and words, demonstrate a strong sense of commitment, control, and challenge, responding to stressful circumstances with an attitude that says stress can be valuable and stressful events at least provide the opportunity to learn and grow.
- *Facilitate positive group sense making of experience.* This relates both to how tasks and missions are planned, discussed, and executed, and also as to how mistakes, failures, and casualties are spoken about and interpreted. For example, do we accept responsibility for mistakes and seek to learn from them or do we blame others and avoid responsibility (and learning)? Leaders build resilience by setting high standards and addressing shortfalls and failures as opportunities to learn and improve. While most of this "sense-making" influence occurs through normal day-to-day interactions and communications, it can also happen in the context of more formal "after-action reviews," or debriefings that focus attention on events as learning opportunities, and create shared positive constructions of events and responses around events.[1]
- *Provide meaningful and challenging group tasks* and then capitalize on group accomplishments by providing recognition, awards, and opportunities to reflect on and magnify positive results (e.g., photographs, news accounts, and other tangible mementos).
- *Communicate a high level of respect and commitment for unit members through both personal examples and policies.* This fosters a strong

---

1. A National Institute of Mental Health (2002) report on best practices for early psychological interventions following mass violence events noted great confusion regarding the term "debriefing." The authors recommend the term "debriefing" be reserved for operational after-action reviews, and not be applied to psychological treatment interventions such as Critical Incident Stress Debriefing. For groups such as the military, after-action group debriefings, properly timed and conducted and focused primarily on events (what happened) rather than emotions and reactions, can have substantial therapeutic value for many participants by helping them to place potentially traumatizing events in a broader context of positive meaning (Bartone, 1997).

sense of commitment among team members to the surrounding social world, or mitwelt, to the organization and one's coworkers.

• Anticipate when high stress events or periods are coming, such as deployments and sustained operations, and spend time beforehand to reinforce mental hardiness among subordinates and junior leaders by sharing experiences, imparting sense-making skills, and focusing on organizational cohesion. Provide as much information as possible to workers as to what they can expect once out on a mission, whether that be an overseas deployment or a long-term duty on an offshore oil drilling platform.

Recommendations for Organizations

Although leadership is important, multiple other factors also influence how individuals make sense of experiences, as well as what leaders can realistically do in this regard. Organizational policies and regulations can not only increase or decrease stress levels, but may also influence commitment, control, and challenge aspects of the hardy-resilient response pattern. Organizations wishing to increase resilience should consider the following:

• *Steps should be taken to identify and select high-hardy leaders as well as provide optimal reinforcement of hardiness in all leaders,* considering the importance of leadership in reducing stress among military personnel of all services. The result should be leaders who understand how to better maintain their own personal mental hardiness as well as how to enhance the hardiness of their subordinates for handling operational stress, especially at lower organizational levels where operational stress may be a more pervasive factor.

• *Leader training and education programs should be reviewed* to determine how they contribute to leaders' resilience under the stress of complex operations. If there are gaps in this area, they should be addressed by devoting greater resources to education and training for psychological resilience.

• *Organizations should develop policies that emphasize the importance of mental hardiness in leaders as well as those being led.* Such policies would assist current leaders at all levels in maintaining the hardiness of their units to be resilient under stress. In the military organization, this would be most critical for forces engaged in land operations-mainly the Army, Marines and Special Operations Forces (SOF). Command information programs (corporate communications, newsletters, internal television programs, webcasts etc) are one means to rapidly address this key topic. Awards, honors and public recognition are other tools leaders can use to reinforce commitment and hardiness.

CONCLUSION

High stress, high risk occupations need resilient workers who can cope effectively with a range of job stressors and not allow stress to degrade their health or job performance. Based on extensive research, we identify psychological hardiness and its key components–commitment, control, and challenge–as the main factors contributing to stress resilience for individuals.

This chapter has applied the psychological hardiness-resilience framework to the high stress occupation of the military as an illustrative case study. We have tried to show how, through leader actions and organizational policies, hardiness-resilience in the entire workforce can be increased to more optimal levels. The same concepts regarding building and sustaining resilience can be applied to many other occupations and organizations where the stressors are multiple and often extreme and the costs of failure severe.

Note: This chapter draws on an earlier report: Bartone, P. T., Barry, C. L., & Armstrong, R. E. (2009 November). *To build resilience: Leader influence on mental hardiness* (Defense Horizons # 69). Washington, DC: Center for Technology and National Security Policy, National Defense University.

REFERENCES

Alford, J. D., & Cuomo, S. A. (2009). Operational design for ISAF in Afghanistan: A primer. *Joint Force Quarterly, 53,* 92–98.

Allport, G. W. (1985). The historical background of social psychology. In G. Lindzey & E. Aronson (Eds.), *Handbook of social psychology* (Vol. 1, 3rd ed., pp. 1-46). New York: Random House.

Azar, I. A. S., Vasudeva, P., & Abdollahi, A. (2006). Relationship between quality of life, hardiness, self-efficacy and self-esteem amongst employed and unemployed married women in Zabol. *Iran Journal of Psychiatry, 1,* 104–111.

Bartone, P. T. (1989). Predictors of stress-related illness in city bus drivers. *Journal of Occupational Medicine, 31,* 657–663.

Bartone, P. T. (1993, June). *Psychosocial predictors of soldier adjustment to combat stress.* Paper presented at the Third European Conference on Traumatic Stress, Bergen, Norway.

Bartone, P. T. (1995, August). *A short hardiness scale.* Paper presented at American Psychological Society annual convention, New York.

Bartone, P. T. (1996). *Stress and hardiness in U.S. peacekeeping soldiers.* Paper presented at the American Psychological Association Annual Convention, Toronto.

Bartone, P. T. (1997, June). *Predictors and moderators of PTSD in American Bosnia forces.* Paper presented at the Fifth European Conference on Traumatic Stress, Maastricht, the Netherlands.

Bartone, P. T. (1999). Hardiness protects against war-related stress in Army reserve forces. *Consulting Psychology Journal, 51,* 72–82.

Bartone, P. T. (2000). Hardiness as a resiliency factor for United States forces in the Gulf War. In J. M. Violanti, D. Paton, & C. Dunning (Eds.), *Posttraumatic stress intervention: Challenges, issues, and perspectives* (pp. 115–133). Springfield, IL: Charles C Thomas.

Bartone, P. T. (2007). Test-retest reliability of the Dispositional Resilience Scale-15, a brief hardiness scale. *Psychological Reports, 101,* 943–944.

Bartone, P. T., Adler, A. B., & Vaitkus, M. A. (1998). Dimensions of psychological stress in peacekeeping operations. *Military Medicine, 163,* 587–593.

Bartone, P. T., Johnsen, B. H., Eid, J., Brun, W., & Laberg, J. C. (2002). Factors influencing small unit cohesion in Norwegian Navy officer cadets. *Military Psychology, 14,* 1–22.

Bartone, P. T., Roland, R., Picano, J. J., & Williams, T. J. (2008). Personality hardiness predicts success in U.S. Army Special Forces candidates. *International Journal of Selection and Assessment, 16,* 78–81.

Bartone, P. T., Ursano, R. J., Wright. K. W., & Ingraham, L. H. (1989). The impact of a military air disaster on the health of assistance workers: A prospective study. *Journal of Nervous and Mental Disease, 177,* 317–328.

Bell, D. B., Bartone, J., Bartone, P. T., Schumm, W. R., & Gade, P. A. (1997). *USAREUR family support during Operation Joint Endeavor: Summary report* (ARI Special Report 34). Alexandria, VA: U.S. Army Research Institute for the Behavioral and Social Sciences. ADA339016.

Berger, P. L., & Luckmann, T. (1966). *The social construction of reality.* Garden City, NY: Doubleday.

Binswanger, L. (1963). *Being in the world: Selected papers of Ludwig Binswanger.* New York: Basic Books.

Bonanno, G. A. (2004). Loss, trauma and human resilience: Have we underestimated the human capacity to thrive after extremely aversive events? *American Psychologist, 59,* 20–28.

Britt, T. W., Adler, A. B., & Bartone, P. T. (2001). Deriving benefits from stressful events: The role of engagement in meaningful work and hardiness. *Journal of Occupational Health Psychology, 6,* 53–63.

Castro, C., & Adler, A. (1999). OPTEMPO: Effects on soldier and unit readiness. *Parameters, 29,* 86–95.

Contrada, R. J. (1989). Type A behavior, personality hardiness, and cardiovascular responses to stress. *Journal of Personality and Social Psychology, 57,* 895–903.

Florian, V., Mikulincer, M., & Taubman, O. (1995). Does hardiness contribute to mental health during a stressful real life situation? The role of appraisal and coping. *Journal of Personality and Social Psychology, 68,* 687–695.

Frankl, V. (1960). *The doctor and the soul.* New York: Knopf.

Funk, S. C. (1992). Hardiness: A review of theory and research. *Health Psychology, 11,* 335–345.

Gal, R. (1987). Military leadership for the 1990s: Commitment-derived leadership. In L. Atwater & R. Penn (Eds.), *Military leadership: Traditions and future trends.* Annapolis, MD: U.S. Naval Academy.

Hancock, P. A., & Krueger, G. P. (2010). Hours of boredom, moments of terror: Tempered desynchrony in military and security force operations. National Defense University Defense & Technology Paper No. 78. Washington, DC: National Defense University Press.

Heidegger, M. (1986). *Being and time.* New York: HarperCollins Publishers.

Hoge, C. W., Castro, C. A., Messer, S. C., McGurk, D., Cotting, D. I., & Koffman, R. L. (2004). Combat duty in Iraq and Afghanistan, mental health problems, and barriers to care. *New England Journal of Medicine, 351,* 13–22.

Hystad, S. W., Eid, J., Johnsen, B. H., Laberg, J. C., & Bartone, P. T. (2010). Psychometric properties of the revised Norwegian dispositional resilience (hardiness) scale. *Scandinavian Journal of Psychology, 51,* 237–245.

Janis, I. (1982). *Groupthink* (2nd ed.). Boston: Houghton Mifflin.

Kobasa, S. C. (1979). Stressful life events, personality, and health: An inquiry into hardiness. *Journal of Personality and Social Psychology. 37,* 1–11.

Kobasa, S. C., Maddi, S. R., & Kahn, S. (1982) Hardiness and health: A prospective study. *Journal of Personality & Social Psychology, 42,* 168–177.

Krueger, G. P. (2008). Contemporary and future battlefields: Soldier stresses and performance. In P. A. Hancock & J. L. Szalma (Eds.), *Performance under stress* (pp. 19–44). Aldershot, England: Ashgate Publishing, Ltd.

Krulak, C. C. (1999, January). The strategic corporal: Leadership in the Three Block War. *Marine Magazine, 28,* 28–34.

Maddi, S. R. (1967). The existential neurosis. *Journal of Abnormal Psychology, 72,* 311–325.

Maddi, S. R. (1987). Hardiness training at Illinois Bell Telephone. In J. P. Opatz (Ed.), *Health promotion evaluation.* Stephens Point, WI: National Wellness Institute.

Maddi, S. R., & Kobasa, S. C. (1984). *The hardy executive.* Homewood, IL: Dow Jones-Irwin.

Maddi, S. R., Harvey, R., Khoshaba, D., Fazel, M., & Resurreccion, N. (2009). Hardiness training facilitates performance in college. *Journal of Positive Psychology, 6,* 566–577.

Marlowe, D. H. (1986). The human dimensions of battle and combat breakdown. In R. Gabriel (Ed.), *Military psychiatry* (pp. 7–24). Westport, CT: Greenwood Press.

National Institute of Mental Health. (2002). *Mental health and mass violence: Evidence-based early psychological intervention for victims/survivors of mass violence* [A workshop to reach consensus on best practices]. (NIH Publication No. 02-5138). Washington, DC: U.S. Government Printing Office. Available at http://www.nimh.nih.gov/health/publications/massviolence.pdf

op den Buijs, T., Andres, M., & Bartone, P. T. (2010). Managing the well-being of military personnel and their families. In J. Soeters, P. C. van Fenema, & R. Beeres (Eds.), *Managing military organizations: Theory and practice* (pp. 240–254). London, Routledge.

Roth, D. L., Wiebe, D. J., Fillingim, R. B., & Shay, K. A. (1989). Life events, fitness, hardiness, and health: A simultaneous analysis of proposed stress-resistance effects. *Journal of Personality and Social Psychology, 57,* 136–142.

Watson, P. J., Ritchie, E. C., Demer, J., Bartone, P., & Pfefferbaum, B. J. (2006). Improving resilience trajectories following mass violence and disaster. In E. C. Ritchie, P. J. Watson, & M. J. Friedman (Eds.), *Interventions following mass violence and disasters: Strategies for mental health practice.* New York: Guilford.

Weick, K. E. (1995). *Sensemaking in organizations.* Thousand Oaks, CA: Sage.

Weisaeth, L., & Sund, A. (1982). Psychiatric problems in UNIFIL and the U.N. soldier's stress syndrome. *International Review of the Army, Navy and Air Force Medical Services, 55,* 109–116.

Westman, M. (1990). The relationship between stress and performance: The moderating effect of hardiness. *Human Performance, 3,* 141–155.

Wiebe, D. J. (1991). Hardiness and stress moderation: A test of proposed mechanisms. *Journal of Personality and Social Psychology, 60,* 89–99.

Zach, S., Raviv, S., & Inbar, R. (2007). The benefits of a graduated training program for security officers on physical performance in stressful situations. *International Journal of Stress Management, 14,* 350–369.

Chapter 9

AN ECOLOGICAL THEORY OF RESILIENCE AND ADAPTIVE CAPACITY IN EMERGENCY SERVICES

Douglas Paton, John M. Violanti, Kim Norris, and Tegan Johnson

INTRODUCTION

Protective and emergency services officers (e.g., police officers, fire fighters) are regularly exposed to critical incidents. Experience of critical incidents can be a precursor to the development of acute and chronic posttraumatic stress reactions. This is not, however, the only outcome that can occur. Critical incident experience can also preface the occurrence of adaptive (e.g., posttraumatic growth, enhanced sense of professional efficacy) outcomes (Aldwin, Levenson, & Spiro, 1994; Armeli, Gunthert, & Cohen, 2001; Joseph & Linley, 2005; Paton, Violanti, Burke, & Gherke, 2009; Paton, Violanti, & Smith, 2003). That pathological outcomes can no longer be regarded as a fait accompli of critical incident exposure and growing evidence for resilient, adaptive, and growth outcomes makes identifying the predictors of these more salutary outcomes an important research goal. This is not to deny the potential for emergency professionals to experience pathological outcomes (e.g., posttraumatic stress disorder [PTSD]), they can (Figure 9.1). However, if the processes and competencies that predict resilient and adaptive outcomes can be identified, this knowledge can inform the proactive development of critical incident stress risk management strategies. Before exploring how salutary outcomes might ensue for critical incident experience, some clarification of the terminology used to describe the various outcomes is required.

The objective of this chapter is to discuss how organizational and family dynamics can contribute to the ability of emergency and protective services officers and agencies to pursue the goal of coping with and adapting to challenging experiences in ways that progressively develop their capacity to deal with future critical incidents. The perspective adopted here encompasses both the ability to cope with (resilience) a given event, but adds a learning component that includes the capacity to mobilize past experiences in ways that develop one's capacity to deal with future events (adaptation). It is important to differentiate between these terms.

RESILIENCE AND ADAPTIVE CAPACITY

The term "resilience" implies a sense of being able to "bounce back" following a challenging experience. This usage reflects the derivation of resilience from its Latin root, "resiliere," meaning "to jump back." This outcome equates to being able to cope with or assimilate (Joseph & Linley, 2005) a new, challenging experience using one's existing mental models (Figure 9.1). If it were possible to predict what emergency professionals would encounter in their operational life, the concept of resilience and the capacity to cope and assimilate it confers on officers would encompass all eventualities. However, this is not an accurate depiction of the experience of emergency professionals.

The nature of contemporary emergency and protective services work is characterized by uncertainty with regard to the nature, complexity, and duration of the events officers could encounter in the future (Paton & Violanti, 2007). This unpredictability introduces a need to include a proactive learning component in the conceptualization of risk management. Accordingly, this chapter argues for a need to include the concept adaptive capacity (Klein, Nicholls, & Thomalla, 2003) when conceptualizing risk management. This concept corresponds to Linley and Joseph's (2005) discussion of the relationship between psychological accommodation (i.e., novel experiences contribute to the development of new, more sophisticated mental models)

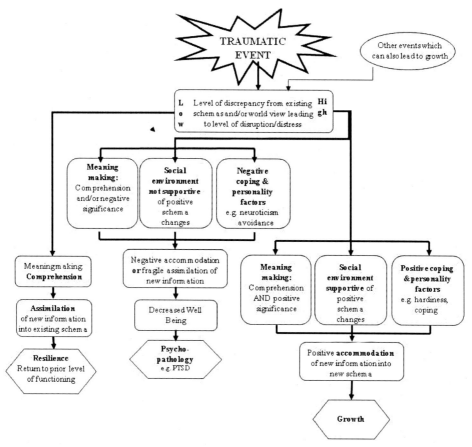

Figure 9.1. The process of assimilation and accommodation following challenging experiences. Adapted from Joseph and Linley (2005).

and adversarial growth (Figure 9.1). Drawing on these perspectives, this chapter is concerned with identifying the individual, team, and agency resources and competencies that officers and agencies can draw on to help them assimilate (resilience) and adapt to, develop, or grow (adaptive capacity) from the demands, challenges, and changes encountered during and after a critical incident.

Critical incidents create a sense of psychological disequilibrium that results when the interpretive frameworks or schemata that guide officers' expectations and actions have lost their capacity to organize novel experiences in meaningful and manageable ways (Dunning, 2003; Janoff-Bulman, 1992; Paton, 1994). A person's ability to deal with challenging circumstances and their capacity to sustain their well-being reflects the extent to which they can use their psychological and physical resources and competencies in ways that allow them to render challenging events coherent, manageable, and meaningful (Antonovsky, 1990; Dunning, 2003).

The goal of critical incident risk management thus becomes one of identifying the resources and processes that facilitate officers' ability to broaden the range of (unpredictable) experiences they can render coherent, meaningful, and manageable (Frederickson, Tugade, Waugh & Larkin, 2003; Paton, 1994, 2006)–that is, to increase the range of future experiences that they will be able to impose meaning on. In order to identify the interpretive processes that contribute to officers' capacity to assimilate and accommodate critical incident experiences, it is first necessary to assess officers' experiences. This poses a challenge.

Critical incidents, and officers' experience of them, is characterized by considerable diversity. For example, officers can encounter critical incidents that could range from a mass casualty road traffic incident to a natural disaster to dealing with the consequences of a dirty bomb detonation in a terrorist attack. Furthermore, officers will have experienced several critical incidents over time, and, depending on the nature of their experience and how they processed it, this would have resulted in officers having accumulated experience of positive and negative outcomes. It is the accumulative nature of critical incident experience that poses the problem.

Most research on critical incidence stress occurs at a specific point in time and usually after a specific incident. When assessing the implications of a given event, it is important to realize that what is being assessed is a product of both the experience of that recent event and the officer's accumulated experiences of previous events and outcomes (Paton & Smith, 1999). The fact that assessment encompasses past and present experience highlights the need to develop a way to capture the diversity of experiences and positive and negative outcomes in ways that encompass meaningfulness, manageability, and coherence in work contexts. One construct capable of fulfilling this goal is job satisfaction.

SATISFACTION AND RESILIENCE

Thomas and Tymon (1994) found a relationship between perceptions of meaning found in work tasks (meaningfulness) and enhanced job satisfaction. Spreitzer, Kizilos, and Nason (1997) observed a positive relationship between competence (manageability) and job satisfaction. These findings have been echoed in the critical incident literature, with finding meaning and benefit (coherence) in emergency work being manifest in changes in levels of job satisfaction (Britt, Adler, & Bartone, 2001; North et al., 2002). Furthermore, because officers weigh up positive and negative experiences when reporting job satisfaction (Hart & Cooper, 2001), assessing job satisfaction at any one time affords a way of assessing the cumulative experience of both positive and negative posttrauma outcomes. This confers on the job satisfaction construct a capacity to capture changes in the meaningfulness, coherence, and manageability facets of resilience, as well as officers' cumulative experience of positive and negative outcomes.

As such, job satisfaction represents a proxy measure capable of assessing adaptation. The higher the satisfaction score, the greater the degree to which the experience of meaning, manageability, and coherence will have increased the ratio of positive to negative outcomes. Having identified a means by which adaptation and resilience can be assessed, the next task is to identify their precursors—that is, to identify the interpretive mechanisms and processes used to adapt.

AN ECOLOGICAL APPROACH

Building on and integrating recent empirical research into how protective services officers adapt to highly challenging circumstances (Burke & Paton, 2006; Johnston & Paton, 2003; Paton, Smith, Violanti & Eränen, 2000), this chapter proposes that effective critical incident stress risk management should be conceptualized as an ecological process. This means that not only does a model have to include critical incident experience; it also must explain how critical incident experience interacts with interpretive processes at person, team, organizational, and family levels to explain how adaptive capacity develops and is sustained.

In particular, the adoption of an ecological perspective argues for greater attention to be given to the family and organizational levels of analysis than has hitherto been the case. Discussion commences with considering why the family level of analysis should be considered as fundamental to a comprehensive theory.

Family

In investigating posttrauma outcomes in seamen involved in rescue efforts following the sinking of the Estonia ferry in 1994 (Eränen, Millar, & Paton, 2000; Paton et al., 2000), analysis of posttrauma outcomes revealed that for those who received psychological debriefing, the effectiveness of this intervention was mediated by changes in social support quality and changes in family functioning (Figure 9.2). Thus, finding debriefing to be effective was translated into higher quality social support skills (e.g., increased emotional disclosure, ability to seek and use support from others). Improved social competence, in turn, enriched the quality of family functioning and facilitated the ability of the family experience to influence on posttrauma outcome (Figure 9.2). Thus, the family became a resource that assisted assimilation of experience and may have contributed to the development of adaptive capacity. A role for social skills and family relationships in trauma mitigation was also reported by Scotti et al. (1995) and Wraith (1994).

Two things can be inferred from these findings. This first reiterates the importance of social support as a recovery resource and the need for the inclusion of social or peer support in critical incident risk management. Organizations have the capacity to offer not only peer support but also training to enhance the intrapersonal and social competencies (e.g., facilitate emotional disclosure, develop social skills) required to use it effectively (Paton, 1997; Scotti et al., 1995). Doing so can, in turn, lead to these competencies spilling from the organizational environment into the family domain. This introduces the second, and more novel, finding concerning the relatively prominent role played by family functioning as a significant predictor of posttrauma outcomes. Family influence is rarely considered in the context of primary critical incident stress prevention programs, but it clearly should be. The importance of doing so can be discerned in another aspect of the work-family relationship—shift work.

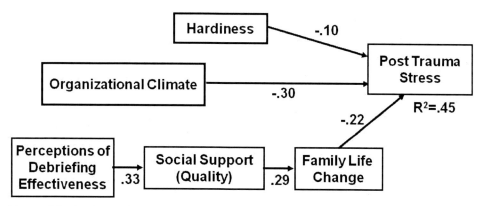

Figure 9.2. Analysis of posttrauma outcomes in Estonia ferry rescue workers. Adapted from Paton et al. (2000).

Shift work, a common component of emergency service work, has a well-documented history as an example of an organizational practice capable of influencing family functioning (Shakespeare-Finch, Paton, & Violanti, 2003). The number and timing of hours worked outside the home significantly influences workers' ability to participate in and enjoy family life. It can thus affect both the quantity and quality of interaction that officers enjoy in family contexts. When seeking explanation for this, it is often assumed that the traumatic nature of officers' experiences has a greater influence on family relationships than organizational factors such as shift work. This assumption is incorrect (Wraith, 1994).

Shakespeare-Finch and colleagues (2003) discussed how, when shift work was controlled for, there was no difference in family functioning between an emergency services shift-work group and a control group comprised of nonemergency shift workers. This suggests that shift work, rather than officers' traumatic experience per se, contributes to family problems. If this finding is integrated with those from the Estonia study (see above), it is evident that organizational choices (e.g., regarding family-friendly shift-work policies, family support groups, involving family members in recovery planning) can affect the quality of officers' relationship with their families and consequently influence the capacity of the family to act as a resilience resource. Thus, decisions that emergency service agencies make regarding elements of the work-family relationship can contribute to the proactive management of stress risk.

Organizations also have a direct influence on how officers render critical incident experience meaningful and coherent. The argument for giving the organization a pivotal role in this process derives from how experience in and of organizations defines the context within which officers experience and interpret critical incidents and their sequelae and within which future capabilities are nurtured or restricted (Paton et al., 2009).

Organizational Influences

Officers respond to incidents as members of an agency whose organizational culture influences their thoughts and actions and represents the context in which challenging experiences (using schema whose nature derives from patterns of interaction with colleagues, senior officers, and organizational procedures over time) are made sense of (Paton, Smith, Ramsay, & Akande, 1999; Paton et al., 2009; Weick & Sutcliffe, 2007). In the context of arguing that organizational characteristics be afforded a prominent position as a predictor of resilience, it is necessary to provide evidence to support this view. Evidence is offered here from two perspectives. The first discusses how organizational life influences officers' interpretation of traumatic experiences. The second explores how perceptions of organizational characteristics can have a direct effect on posttrauma outcomes.

Paton et al. (1999) explored how organizational experience influences traumatic reactivity by examining the psychometric structure of responses to traumatic events assessed using the Impact of Event Scale (IES) (Horowitz et al., 1979). If a scale is accessing a latent construct that is a robust measure of the experience of traumatic stress, the structural relations between items should hold irrespective of, for example, the professional or organizational membership of those being tested. To test this using the IES, Paton and colleagues compared the structural dimensions of traumatic reactivity across organizations (e.g., human service and emergency service organizations), professions (e.g., fire fighters vs. social workers), and countries (e.g., by comparing Scottish, Nigerian, and Australian firefighters).

Multidimensional scaling was used to examine structural relationships to determine whether organizational and cultural factors influenced the structure of traumatic stress reactions. It was hypothesized that if the environment does not influence response, a high level of structural homogeneity would be observed across the populations sampled. This hypothesis was not supported. The results indicated that cultural and professional membership did influence how officers experienced traumatic stress reactions (Paton et al., 1999). Because of its ability to influence how critical incident experiences are interpreted and made sense of, a prime source of this organizational influence is organizational culture (Paton et al., 2009; Weick & Sutcliffe, 2007).

ORGANIZATIONAL CULTURE

In the study of the rescue workers involved in the response to the Estonia ferry sinking (see above), Paton et al. (2000) demonstrated that perceptions of organizational culture was the best predictor of critical incident stress (Figure 9.2). A role for organizational characteristics was also evident in Hart and Cooper's (2001) conceptual model of organizational health, which predicted that interaction between individual and organizational factors influences well-being. Of particular relevance for the present chapter was the central role that Hart and Cooper afforded organizational climate (officers' perceptions of how their organization functions and how these perceptions influence both their well-being and the performance of their organizational role).

Burke and Paton (2006) tested the ability of this model to predict satisfaction in the context of emergency responders (police, fire, and paramedic populations) experience of critical incidents. They reported how interaction among organizational culture, officers' experience of daily hassles and uplifts (Hart & Wearing, 1995; Hart, Wearing, ^ Heady, 1993), and officers' problem- and emotion-focused coping style accounted for 44 percent of the variance in job satisfaction. Organizational culture was the best single predictor of job satisfaction and, by inference, the best indicator of officers' ability to render their critical incident experiences coherent, meaningful, and manageable.

Finding a prominent role for organizational characteristics reinforces the need for organizational factors to be afforded a central position in the process of modeling resilience. This does not, however, mean that other intra- and interpersonal variables do not have a role to play. Hart and Cooper (2001) argued that the inclusion of individual (e.g., personality, hardiness) and group (e.g., peer and supervisor support) constructs could help account for additional variance. Hardiness was found to be a predictor of traumatic stress outcomes in the Estonia study (Figure 9.2).

Although the studies discussed in this section identify a role for organizational culture in the traumatic stress process, it remains to be seen how officers' experience in organizational contexts translates into the development of schema or interpretive frameworks (derived and sustained from officer, team, organizational, and family interaction) that can be developed and enacted in ways that facilitate the development of officers' adaptive capacity. One construct that can offer insights into how this might be achieved is empowerment.

EMPOWERMENT

Because it has demonstrated a capacity to predict satisfaction in individuals and teams (Kirkman & Rosen, 1999; Koberg, Boss, Senjem, & Goodman, 1999), empowerment is a strong candidate for acting as a mechanism capable of converting organizational and work experiences into adaptive capacity. The next question involves asking how empowerment can act as a lens through which individual, team, organizational, and family resources are focused in ways that facilitate the development of adaptive capacity and satisfaction. If a sound theoretical rationale for it being able to do so can be demonstrated, this would justify empowerment being included in a model of resilience and adaptive capacity. It is to a discussion of how empowerment satisfies this criterion that this chapter now turns.

Empowerment is a well-used construct in the management literature, usually in relation to processes such as participation and delegation (Conger & Konungo, 1988). Because delegation (e.g., delegating responsibility for crisis decision making) can influence resilience (Paton & Flin, 1999), empowerment can contribute to modeling resilience. However, it is its links to motivating action in conditions of uncertainty (Conger & Konungo, 1988; Spreitzer, 1997) that renders empowerment capable of providing valuable insights into how re-silience and adaptive capacity can be developed and sustained.

Enabling Action

Conger and Konungo (1988) conceptualize empowerment as an enabling process that facilitates the conditions necessary to effectively confront future challenges. Conger and Konungo argue that individual differences in levels of competence are attributable, at least in part, to the degree to which the environment enables actions to take place and can thus be defined as a process that facilitates learned resourcefulness (Johnston & Paton, 2003). If officers have sufficient psychological, social, and physical resources and the capacity to use them, they will be able to effectively confront the challenges presented by events, the environment, and interpersonal relationships (Conger & Konungo, 1988; Spreitzer, 1997). Empowerment may thus facilitate understanding how officers can be enabled to use resources to confront challenging circumstances.

Thomas and Velthouse (1990) echo these sentiments and complement this position by adding that beliefs about future competence derive from the cognitions and interpretive frameworks (developed, in part, through the enabling process of empowerment) that provide meaning to officers' experiences and build their capacity to deal with future challenges. It is the notion of enabling through the development of (empowering) schema that can help explain how officers' experience of organizational culture can be enacted as adaptive and resilience resources.

Conger and Konungo (1988) argue that empowerment describes a process of facilitating competencies using organizational strategies to remove conditions that foster powerlessness (e.g., organizational and operational hassles) and developing formal organizational practices and informal techniques for developing learned resourcefulness (e.g., self-efficacy information and competencies). Placing the development of competencies in an organizational context provides a foundation for the development of organizational strategies to develop and sustain resilience and enables officers to effectively confront future challenges (Johnston & Paton, 2003).

By linking the organizational environment and the schema that underpin future adaptive capacity, empowerment theories have considerable potential to inform understanding of how resilience is enacted in organizational contexts. This capability is further bolstered by the fact that empowerment is conceptualized as an iterative process involving a cycle of environmental events, behaviors and task assessments (Johnston & Paton, 2003). The iterative nature of empowerment makes its construct capable of accommodating the repetitive nature of officers' involvement in critical incidents and the learning process required to maintain adaptive capacity in the changing environment of contemporary emergency services work.

Environmental events (e.g., critical incidents) provide information to officers about both the consequences of their previous task behavior and the conditions they can expect to experience in future task behavior (Conger & Konungo, 1988). This proactive perspective is particularly important for a theory that seeks to identify how future capability can be developed and sustained. In addition to it emanating from their own experiences, information about novel experiences can be provided by peers, subordinates, and superiors at work, in the context of, for example, performance appraisals, training programs, and meetings. This feedback process provides a foundation for developing future adaptive capacity (Paton & Jackson, 2002). Thus, through each progressive cycle of event (e.g., following a challenging critical incident), assessment, and feedback, officers construct the operational schema they use to respond to, plan for, and interpret critical incidents. For it to inform understanding of how officers can accommodate lessons from experiences (Figure 9.1), it is necessary to identify how empowerment cycles inform the development of adaptive capacity. This is accomplished through the way in which the environmental assessment process translates into two outcomes: task assessment and global assessment.

Task Assessment

According to Thomas and Velthouse (1990), officers' task assessments comprise several components. The first, meaningfulness, describes the degree of congruence between the tasks performed and one's values, attitudes, and behaviors. Empowered individuals feel a sense of personal significance resulting from their involvement in work activities. A sense of meaning results in officers experiencing feelings of purpose, energy, and commitment toward their work (Spreitzer, 1997; Thomas & Velthouse, 1990).

The second component, competence, reflects officer's beliefs in their ability to perform successfully in their operational role (Spreitzer, 1997). This outcome is comparable to the concept of manageability (Antonovsky, 1990; Dunning, 2003). Because there is a direct relationship between competence and the effort and persistence invested in confronting significant physical and psychological demands, competence makes an important contribution to officers' capacity to adapt to challenging circumstances.

The third facet, choice, reflects the extent to which officers perceive that their behavior is self-determined (Spreitzer, 1997). A sense of choice is achieved when officers believe they are actively involved in defining how they perform their role (a prominent item when reporting organizational uplifts such as being given responsibility for making decisions), rather than just being passive recipients (as is often the case when describing organizational hassles such as dealing with excessive red tape). This renders this facet of empowerment comparable to the notion of coherence (Antonovsky, 1990; Dunning, 2003). A sense of choice is particularly important for dealing with emergent, contingent emergency demands and for creative and crisis decision making and the development and maintenance of situational awareness when responding to critical incidents (Paton & Flin, 1999). An ability to exercise choice also facilitates learning from training and operational experiences and transmitting the lessons learned to others. The ability to draw direct comparison between the components of task assessment and the concepts of meaningfulness, manageability, and coherence adds weight to the argument that empowerment is a prime candidate as a mechanism that can inform understanding how resilience and adaptive capacity develop in emergency service organizations.

The final task assessment component, impact, encompasses officers' beliefs that they can influence important organizational outcomes (Spreitzer, 1997). Where as choice concerns control over one's work behaviors, impact concerns the notion of personal input into the attainment of organizational outcomes. Parallels can be drawn between this element and perceived control, another factor that has been widely implicated in thinking on stress resilience and adaptability. The final form of assessment, global assessment, further establishes empowerment as a construct that can inform the understanding of officers' resilience and their capacity to adapt to future incidents.

Global Assessments

Whereas task assessments relate to a specific event and time, global assessments embody the capacity to generalize expectancies and learning across tasks and over time. Global assessments describe a capacity to fill in gaps when faced with new and/or unfamiliar situations (Thomas & Velthouse, 1990). This aspect of empowerment is essential for adaptive capacity in professions whose members cannot predict what they will be called on to confront (e.g., the next call-out could be a gas main explosion or the detonation of a dirty bomb) and must be able to use current experiences as a basis for preparing to deal with future risk and uncertainty.

Both global and task assessments, and thus the capacity to adapt, are influenced by officers' interpretive styles, with schema emanating from one of several interpretive styles (Thomas & Velthouse, 1990). According to Thomas and Velthouse (1990), interpretive frameworks are influenced by the work context, with management practices being important influences on how they are developed and sustained.

Interpretive Styles

Two interpretive styles, envisioning and evaluation, facilitate the development of adaptive capacity. Envisioning refers to how officers anticipate or envision future events and outcomes. This schema component influences the kinds of attributions officers make about critical incident experiences. Officers who anticipate positive rather than negative outcomes experience stronger task and global assessments. This, in turn, increases the likelihood of officers approaching future events as learning experiences. Given that officers face the prospect of experiencing novel events or events with novel characteristics and thus inevitably have to confront complex, dynamic incidents, the kind of learning culture that can ensue from creating the environmental conditions to encourage envisioning increases the likelihood that response problems and challenges will be perceived as catalysts for future development and not as failure (Paton, 2006; Paton & Stephens, 1996).

A second schema component, evaluation, refers to how officers evaluate success or failure. Thomas and Velthouse (1990) argue that individuals who adopt less absolutist and more realistic standards experience greater levels of empowerment. This observation is reinforced by findings that officers who approach critical incidents with realistic performance expectations and acknowledge environmental limitations on their outcomes more readily adapt to challenging circumstances (Paton, 1994; Raphael, 1986).

This discussion illustrates how empowerment represents a mechanism that can be invoked to explain how officers' experience of organizational culture is translated into adaptive capacity and resilience. Having done so, the final issue to be tackled is identifying predictors of empowerment that can be incorporated into a comprehensive model.

PREDICTING EMPOWERMENT

Several antecedents of psychological empowerment have been identified (Ripley & Ripley, 1992; Spreitzer, 1995a,b, 1996). Prominent among these are personal characteristics (e.g., personality) and social structural variables (e.g., access to resources and information, organisational trust, peer cohesion, and supervisory support). This literature can contribute to identifying predictors of empowerment that can be incorporated in an ecological theory of critical incident stress risk management.

Personal Characteristics

Several dispositional or intrapersonal characteristics have been identified as having a relationship with empowerment. One of these, hardiness (see Chapter 8), has a long history as a predictor of resilience, and one whose existence is strongly influenced by the officer-agency relationship (Bartone, 2004). It may be an important adjunct to empowerment. Although organizational decisions can provide the conditions necessary to enable officers (i.e., create empowering settings), this cannot be taken to automatically imply that officers will fully utilize these opportunities. It is necessary to complement an enabling environment with empowered officers. The control, challenge, and commitment facets of hardiness describe competencies indicative of officers' potential to utilize environmental opportunities to learn from experiences. A role for hardiness was found in the Estonia study introduced above (Figure 9.1). For this reason, hardiness is included in the model.

Another variable that has attracted interest is the personality dimension of conscientiousness, particularly with regard to its attributes of achievement orientation and dependability (McNaus & Kelly, 1999). Thomas and Velthouse (1990) argue that conscientious individuals typically experience a stronger sense of competence in the tasks they perform, particularly during times of change and disruption (e.g., responding to critical incidents and the need to adapt to unpredictable emergent demands). Conscientious individuals strive to learn and improve in order to overcome new challenges, see themselves as personally responsible for their actions, seek out opportunities for development, and tend to perform above and beyond official role expectations. It is a characteristic that may thus increase the likelihood of adopting envisioning (see above).

Conscientious individuals experience a stronger sense of meaning and competence in their work, particularly during challenging events (Thomas & Velthouse, 1990), demonstrate greater levels of perseverance in these efforts (Behling, 1998), and are more committed to contributing to team efforts (Hough, 1998). This contributes positively to both the level of cooperation with and support for coworkers that people demonstrate in work contexts and to sustaining a cohesive team response to complex events.

Peer Relationships and Team Cohesion

Liden et al. (2000) found that relations with coworkers predicts officers' sense of psychological empowerment, particularly with regard to creating a level of fit in the relationship between an individual's values and attitudes and those of a work role (i.e., it contributes to meaningfulness and coherence) (Major, Kozlowski, Chao, & Gardner, 1995). The improved coordination resulting from a high level of peer cohesion increases the likelihood of officers experiencing more intrinsic value (meaning) in the performance of a task (Mullen & Copper, 1994; Paton & Stephens, 1996; Perry, 1997). It also acts to increase levels of social support provided to coworkers (George & Bettenhausen, 1990).

Members of cohesive work teams are more willing to share their knowledge and skills, so contributing to the development and maintenance of the learning culture. Cohesive networks are less dependent on senior officers for obtaining important resources, as reciprocal peer relationships are an alternative source for such resources (Liden et al., 1997), contributing to a greater sense of self-determination in one's work. Furthermore, when this peer cohesion coexists with cohesive romantic and family relationships, the availability of supportive resources is further expanded, thereby facilitating employees' experiences of self-efficacy, self-determination, and resilience (Norris, 2010) (see also Figure 9.1). Taken together, the social structural variables of senior officer support, peer, and familial cohesion can make a valuable contribution to a theory of resilience and adaptive capacity. For people and teams to function effectively, they need the right resources.

Resource Availability and Utility

Having insufficient, inadequate, or inappropriate resources (physical, social, and informational) to perform response tasks contributes to critical incident stress risk (Carafano, 2003; Paton, 1994). Resources allow individuals to take initiative and enhance their sense of control over environmental challenges (Gist & Mitchell, 1992; Lin, 1998; Paton, 1994). One resource that plays a pivotal role in predicting empowerment is information.

Crisis information management systems capable of providing information in conditions of uncertainty are essential to adaptive capacity in emergency responders (Paton & Flin, 1999). Access to pertinent information in a timely manner plays an important role in creating a sense of purpose and meaning among officers (Conger & Konungo, 1988). It has also been demonstrated to influence meaning making and perceptions of control within Antarctic expeditioners as well as their family/partners, with higher levels of information availability promoting higher levels of well-being in both participant categories (Norris, 2010). However, information itself is not enough. The social context in which information is received is an equally important determinant of empowerment. In this context, one aspect of the agency-officer relationship becomes particularly important, and that concerns trust.

Trust

Trust is a prominent determinant of the effectiveness of interpersonal relationships, group processes, and organizational relationships (Barker & Camarata, 1998; Herriot, Hirch & Reilly, 1998), particularly when individuals face some potential or actual risk (Coleman, 1990). When dealing with critical incidents, agencies and officers alike have to deal with risk and uncertainty, and trust has been identified as a predictor of people's ability to deal with complex, high risk events (Siegrist & Cvetkovich, 2000), particularly when relying on others to provide information or assistance.

Trust influences perception of others' motives, their competence, and the credibility of the information they provide (Earle, 2004; Kramer, 1999). The perceived credibility of information (see Access to Resources section) influences the extent to which the information available contributes to people's sense of meaning and purpose. Individual are more willing to commit to acting cooperatively in high risk situations when they believe those with whom they must collaborate or work under are competent, dependable, likely to act with integrity (in the present and in the future), and to care for their interests (Dirks, 1999). Trust thus plays a pivotal role in empowering officers (Spreitzer & Mishra, 1999). People functioning in trusting, reciprocal relationships are left feeling empowered and more likely to experience meaning in their work. Levels of trust are influenced by the organizational culture (Barker & Camarata, 1998; Siegrist & Cvetkovich, 2000) and the dynamics of interpersonal relationships (Norris, 2010).

Given the role of trust in this model, and because trust reflects important qualities of the social context at work, it becomes pertinent to consider the relationship between empowerment and social structural aspects of the organization. This line of inquiry is supported by the fact that organizations functioning with cultures valuing openness and trust create opportunities for officers to engage in learning and growth, contributing to the development of officers' adaptive capacity (competence) (Barker & Camarata, 1998). This adaptive capacity is further enhanced when organizational openness and communication extend to involve officers' partners and families who are then more trusting and supportive of the organization and employee's engagement in the work role, thereby minimizing tension between work and nonwork roles (Norris, 2010). A group with a major role to play in creating and sustaining a climate of trust are the senior officers who translate organizational culture into the day-to-day values and procedures that sustain the schema officers engage in to plan for and respond to critical incidents.

Senior Officer Support

The actions of senior officers play a central role in developing and sustaining empowering environments (Liden, Wayne, & Sparrow, 2000; Paton & Stephens, 1996). Leadership practices such as positive reinforcement help create an empowering team environment (Manz & Sims, 1996; Paton, 1994). Positive feedback, encouragement, and constructive discussion of response problems and how they can be resolved in the future from both co-workers and senior officers empower employees (Quinn & Spreitzer, 1997). It does so by drawing one's emphasis away from personal weaknesses in a difficult or challenging situation and replacing it with an active approach to anticipating how to exercise control in future. In this way, senior officer behavior contributes to the development of the attributional, envisioning, and evaluative schema components (see above) that are instrumental in translating officers' organizational experiences into resilient beliefs and adaptive capacity.

A high level of managerial support enhances feelings of competence and informs employees that they are allowed and encouraged to exercise choice (self-determination) about how they conduct some of their work (Langfried, 2000). Quality supervisor-subordinate relationships, of which supportive supervisor behavior is a crucial factor (Liden, Sparrow, & Wayne, 1997), creates the conditions necessary for the personal growth of individuals (Cogliser & Schriesheim, 2000), enhancing general feelings of competence (global assessment). Additionally, quality supervisor-subordinate relationships instigate the creation of similar value structures between individuals (Cogliser & Schriesheim, 2000), building shared schema, enabling employees to find increased meaning in their task activities, and contributing to the development of a sense of cohesion between colleagues. Furthermore, drawing additionally on the findings of the Estonia study (Figure 9.2), the development of a cohesive and supportive work culture may enhance the quality of family functioning. This introduces a need to include family in the process of developing an ecological model of resilience and adaptive capacity.

The Work-Family Interface

The work-family interface refers to the interdependence between work and nonwork roles in the individual employee experience. The interrelationships between work and nonwork (i.e., family) roles occurs in a bidirectional manner, such that work-related demands can interfere with family-related responsibilities (Work Interference with Family; e.g. working late precludes spending time with family), and family-related responsibilities can interfere with work-related demands (Family Interference with Work; e.g. caring for a sick child precludes attending work) (Greenhaus, Collins, Singh, & Parasuraman, 1997; Netemeyer, Boles, & McMurrian, 1996).

Norris (2010) identified the importance of the work-family interface in facilitating empowerment in employees. Specifically, Norris argued that in addition to organizational influences, family processes (including the dynamics of an intimate relationship and family-level coping strategies) mediate the degree of resilience demonstrated by the individual in response to the challenge presented–in this context, hazards associated with employment. Of particular importance is the degree of perceived familial involvement in the employment experience (which can be readily facilitated by organizational culture–see above). Higher levels of perceived familial involvement in the employment experience promote greater communication, trust, and availability of resources (at the individual, organization, and family levels) facilitating enhanced capacity for the employee to experience resilient outcomes in response to challenging events. This was evident in the earlier discussion of the Estonia rescue that described how increasing family resources to assist managing critical incident stress led to improved posttrauma outcomes.

In contrast, lower levels of perceived familial involvement can engender a situation in which the individual is forced to balance the competing and incongruent demands of the organization with the family. The result is that the individual's personal resources are depleted, and are therefore not available to be directed toward the challenging event. In turn this reduces the likelihood of resilient and adaptive outcomes for the individual, and may be a contributing factor to employee turnover and/or relationship dissolution.

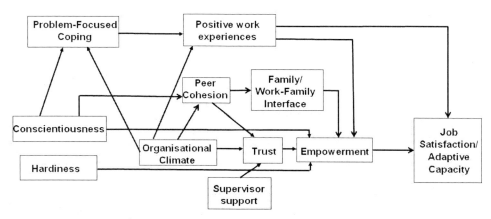

Figure 9.3. The Ecological Model of Adaptive Capacity.

CONCLUSION

Taken together, the work discussed in this chapter suggests that in order to facilitate resilient outcomes in employees, we need to extend our focus beyond the individual-organization relationship to include the individual-organizational-familial relationship. Thus, when developing a comprehensive model of resilience, it is important to incorporate individual-, organizational-, and relationship-level factors that may influence this process. Integrating the perspectives discussed here provides the foundation for the development of an ecological model of resiliency in high risk professions. This model is summarised in Figure 9.3.

Although this model remains to be tested, grounds for it being pursued in practice derive from the fact that it was developed from theoretically robust and empirically tested work. The ecological model describes resilience and adaptive capacity as resulting from the interaction among person, team, organizational, and family factors, with each level of analysis being interdependent. They interact to contribute to the development of empowered officers who exist in empowering settings. This combination of empowered officers and empowering settings supports the development of officers' capacity to learn from and accommodate experiences and to consolidate these outcomes in a set of shared beliefs that constitute adaptive capacity.

All the components, with the exception of conscientiousness, are amenable to change through organizational intervention and change strategies. Guidelines for changing hardiness, peer support, supervisor support, organizational culture, trust, empowerment, and family dynamics are available in the literature (Bartone, 2004; Cogliser & Schriesheim, 2000; Hart et al., 1993; Herriot et al., 1998; Perry, 1997; Quinn & Spreitzer, 1997; Shakespeare-Finch et al., 2003; Wraith, 1994). This confers on the model both theoretical rigor and practical utility.

REFERENCES

Aldwin, C. M., Levenson, M. R., & Spiro, A. III. (1994). Vulnerability and resilience to combat exposure: Can stress have lifelong effects? *Psychology and Aging, 9,* 3444.

Armeli, S., Gunthert, K. C., & Cohen, L. H. (2001). Stressor appraisals, coping, and post-event outcomes: The dimensionality and antecedents of stress-related growth. *Journal of Social and Clinical Psychology, 20,* 366–395.

Antonovsky, A. (1990). Personality and health: Testing the sense of coherence model. In H. S. Friedman (Ed.), *Personality and disease* (pp. 155–177). New York: John Wiley & Sons.

Barker, R. T., & Camarata, M. R. (1998). The role of communication in creating and maintaining a learning organization: Preconditions, indicators, and disciplines. *The Journal of Business Communication, 35,* 443–467.

Behling, O. (1998). Employee selection: Will intelligence and personality do the job? *Academy of Management Executive, 12,* 77–86.

Britt, T. W., Adler, A. B., & Bartone, P. T. (2001). Deriving benefits from stressful events: The role of engagement in meaningful work and hardiness. *Journal of Occupational Health Psychology, 6,* 53–63.

Burke, K., & Paton, D. (2006). Well-being in protective services personnel: Organisational influences. *Australasian Journal of Disaster and Trauma Studies, 2006-2.* Available at http://trauma.massey.ac.nz/issues/2006-2/burke.htm

Carafano, J. J. (2003) *Preparing responders to respond: The challenges to emergency preparedness in the 21st century* (Heritage Lectures #812). Washington, DC: The Heritage Foundation.

Cogliser, C. C., & Schriesheim, C. A. (2000) Exploring work unit context and leader-member exchange: A multi-level perspective. *Journal of Organizational Behaviour, 21,* 487–511.

Coleman, J. S. (1990). *Foundations of social theory.* Cambridge, MA: Belknap Press.

Conger, J. A. (1989). Leadership: The art of empowering others. *The Academy of Management Executive, 3,* 17–24.

Dirks, K. T. (1999) The effects of interpersonal trust on work group performance. *Journal of Applied Psychology, 84,* 445–455.

Dunning, C. (2003) Sense of Coherence in Managing Trauma Workers. In D. Paton, J. M. Violanti, & L. M. Smith (Eds.), *Promoting capabilities to manage posttraumatic stress: Perspectives on resilience.* Springfield, IL: Charles C Thomas.

Earle, T. C. (2004). Thinking aloud about trust: A protocol analysis of trust in risk management. *Risk Analysis, 24,* 169–183.

Eränen, L., Millar, M., & Paton, D. (2000, June). *Organisational recovery from disaster: Traumatic response within voluntary disaster workers.* Paper presented at the International Society for Stress Studies Conference, Istanbul, Turkey.

Fredrickson, B. L., Tugade, M. M., Waugh, C. E. & Larkin, G. (2003). What good are positive emotions in crises?: A prospective study of resilience and emotions following the terrorist attacks on the United States on September 11th, 2001. *Journal of Personality and Social Psychology 84,* 365–376.

Gist, M., & Mitchell, T. N. (1992) Self-Efficacy: A theoretical analysis of its determinants and malleability. *Academy of Management Review, 17,* 183–211.

Greenhaus, J., Collins, K., Singh, R., & Parasuraman, S. (1997). Work and family influences on departure from public accounting. *Journal of Vocational Behavior, 50,* 249–270.

Hart, P. M., & Cooper, C. L. (2001) Occupational stress: Toward a more integrated framework. In N. Anderson, D. S. Ones, H. K. Sinangil, & C. Viswesvaren (Eds.), *International handbook of work and organizational psychology: Vol. 2. Organizational psychology.* London: Sage Publications.

Hart, P. M., & Wearing, A. J. (1995). Occupational stress and well-being: A systematic approach to research, policy and practice. In P. Cotton (Ed.), *Psychological health in the workplace.* Carlton: Australian Psychological Society.

Hart, P. M., Wearing, A. J., & Headey, B. (1993). Assessing police work experiences: Development of the Police Daily Hassles and Uplifts Scales. *Journal of Criminal Justice, 21,* 553–572.

Herriot, P., Hirsh, W., & Reilly, P. (1998). *Trust and transition: Managing today's employment relationship.* Chichester: John Wiley & Sons.

Horowitz, M., Wilner, M., & Alvarez, W. (1979). Impact of Event Scale: A measure of subjective stress. *Psychosomatic Medicine, 41,* 209–218.

Hough, L. M. (1998). Personality at work: Issues and evidence. In M. D. Hakel (Ed.), *Beyond multiple choice: Evaluating alternatives to traditional testing for selection* (pp. 131–166). Mahwah, NJ: Lawrence Erlbaum Associates.

Janoff-Bulman, R. (1992). *Shattered assumptions.* New York: Free Press.

Johnston, P., & Paton, D. (2003) Environmental resilience: Psychological empowerment in high-risk professions. In D. Paton, J. Violanti, & L. Smith (Eds.), *Promoting capabilities to manage posttraumatic stress: Perspectives on resilience.* Springfield, IL: Charles C Thomas.

Joseph, S., & Linley, A.P. (2005). Positive adjustment to threatening events: An organismic valuing theory of growth through adversity. *Review of General Psychology, 9,* 262–280.

Kirkman, B. L., & Rosen, B. (1999). Beyond self-management: Antecedents and consequences of team empowerment. *Academy of Management Journal, 42,* 58–74.

Klein, R., Nicholls, R., & Thomalla, F. (2003). Resilience to natural hazards: How useful is this concept? *Environmental Hazards, 5,* 35–45.

Koberg, C. S., Boss, R. W., Senjem, J. S., & Goodman, E. A. (1999). Antecedents and outcomes of empowerment. *Group & Organization Management, 24,* 71–91.

Kramer, R. M. (1999.) Trust and distrust in organizations: Emerging perspectives, enduring questions. *Annual Review of Psychology, 50,* 569–598.

Langfried, C. W. (2000). The paradox of self-management: Individual and group autonomy in work groups. *Journal of Organizational Behaviour, 21,* 563–585.

Liden, R. C., Sparrow, R. T., & Wayne, S. J. (1997). Leader-member exchange theory: The past and potential for the future. *Research in Personnel and Human Resources Management, 15,* 47–119.

Liden, R. C., Wayne, S. J., & Sparrow, R. T. (2000). An examination of the mediating role of psychological empowerment on the relations between the job, interpersonal relationships, and work outcomes. *Journal of Applied Psychology, 85,* 407–416.

Lin, C. Y. (1998). The essence of empowerment: A conceptual model and a case illustration. *Journal of Applied Management Studies, 7,* 223–238.

Major, D. A., Kozlowski, S. W., Chao, G. T., & Gardner, P. D (1995). A longitudinal investigation of newcomer expectations, early socialization outcomes, and the moderating effects of the role development factors. *Journal of Applied Psychology, 80,* 418–431.

McNaus, M. A., & Kelly, M. L. (1999). Personality measures and biodata: Evidence predicting their incremental predictive value in the life insurance industry. *Personnel Psycholohy, 52,* 137–148.

Mullen, B., & Copper, C. (1994). The relation between group cohesiveness and performance: An integration. *Psychological Bulletin, 115,* 210–227.

Netemeyer, R. G., Boles, J. S., & McMurrian, R. (1996). Development and validation of work-family conflict and family-work conflict scales. *Journal of Applied Psychology, 81,* 400–410.

Norris, K. (2010). *Breaking the ice: Developing a model of expeditioner and partner adaptation to Antarctic employment.* Unpublished doctoral dissertation, University of Tasmania, Australia.

North, C. S., Tivis, L., McMillen, J. C., Pfefferbaum, B., Cox, J., Spitznagel, E. L., Bunch, K., Schorr, J., & Smith, E. M. (2002). Coping, functioning, and adjustment of rescue workers after the Oklahoma City bombing. *Journal of Traumatic Stress, 15,* 171–175.

Paton, D. (1994). Disaster relief work: An assessment of training effectiveness. *Journal of Traumatic Stress, 7,* 275–288.

Paton, D. (2006). Posttraumatic growth in emergency professionals. In. L. Calhoun & R. Tedeschi (Eds.), *Handbook of posttraumatic growth: Research and practice.* Mahwah, NJ: Lawrence Erlbaum Associates.

Paton, D., & Flin, R. (1999). Disaster stress: An emergency management perspective. *Disaster Prevention and Management, 8,* 261–267.

Paton, D., & Jackson, D. (2002). Developing disaster management capability: An assessment centre approach. *Disaster Prevention and Management, 11,* 115–122.

Paton, D., & Smith, L.M. (1999). Assessment, conceptual and methodological issues in researching traumatic stress in police officers. In J. M. Violanti & D. Paton (Eds.), *Police trauma: Psychological aftermath of civilian combat.* Springfield, IL, Charles C Thomas.

Paton, D., Smith, L. M., Ramsay, R., & Akande, D. (1999). A structural re-assessment of the Impact of Event Scale: The influence of occupational and cultural contexts. In R. Gist & B. Lubin (Eds.), *Response to disaster.* Philadelphia, PA: Taylor & Francis.

Paton, D., Smith, L. M., Violanti, J., & Eranen, L. (2000). Work-related traumatic stress: Risk, vulnerability and resilience. In D. Paton, J. M. Violanti, & C. Dunning (Eds.), *Posttraumatic stress intervention: Challenges, issues and perspectives* (pp. 187–204). Springfield, IL: Charles C Thomas.

Paton, D., & Stephens, C. (1996). Training and support for emergency responders. In D. Paton & J. Violanti (Eds.), *Traumatic stress in critical occupations: Recognition, consequences and treatment.* Springfield, IL: Charles C Thomas.

Paton, D., & Violanti, J. M. (2007). Terrorism stress risk assessment and management. In B. Bonger, L. Beutler, & P Zimbardo (Eds.), *Psychology of terrorism.* San Francisco: Oxford University Press.

Paton, D., Violanti, J. M., Burke, K., & Gherke, A. (2009). *Traumatic stress in police officers: A career length assessment from recruitment to retirement.* Springfield, IL: Charles C Thomas.

Paton, D., Violanti, J. M., & Smith, L. M. (2003). *Promoting capabilities to manage posttraumatic stress: Perspectives on resilience.* Springfield, IL: Charles C Thomas.

Perry, I. (1997). Creating and empowering effective work teams. *Management Services, 41*, 8–11.

Quinn, R. E., & Spreitzer, G. M. (1997). The road to empowerment: Seven questions every leader should consider. *Organisational Dynamics, Autumn,* pp. 7-49.

Raphael, B. (1986). *When disaster strikes.* London: Hutchinson.

Ripley, R. E., & Ripley, M. J. (1992). Empowerment, the cornerstone of quality: Empowering management in innovative organisations in the 1990's. *Management Decision, 30,* 20–43.

Scotti, J. R., Beach, B. K., Northrop, L. M. E. Rode, C. A., & Forsyth, J. P. (1995). The psychological impact of accidental injury. In J. R. Freedy & S. E. Hobfoll (Eds.), *Traumatic stress: From theory to practice.* New York: Plenum Press.

Shakespeare-Finch, J., Paton, D., & Violanti, J. (2003). The family: Resilience resource and resilience needs. In D. Paton, J. Violanti, & L. Smith (Eds.), *Promoting capabilities to manage posttraumatic stress: Perspectives on resilience.* Springfield, IL: Charles C Thomas.

Siegrist, M., & Cvetkovich, G. (2000). Perception of hazards: The role of social trust and knowledge. *Risk Analysis, 20,* 713–719.

Spreitzer, G. M. (1995a). An empirical test of a comprehensive model of intrapersonal empowerment in the workplace. *American Journal of Community Psychology, 23,* 601–629.

Spreitzer, G. M. (1995b). Psychological empowerment in the workplace: Dimensions, measurement, and validation. *Academy of Management Journal, 38,* 1442–1465.

Spreitzer, G. M. (1996). Social structural characteristics of psychological empowerment. *Academy of Management Journal, 39,* 483–504.

Spreitzer, G. M. (1997). Toward a common ground in defining empowerment. *Research in Organizational Change and Development, 10,* 31–62.

Spreitzer, G. M., Kizilos, M. A., & Nason, S. W. (1997). A dimensional analysis of the relationship between psychological empowerment and effectiveness, satisfaction and strain. *Journal of Management, 23,* 679–704.

Spreitzer, G. M., & Mishra, A. K. (1999). Giving up control without losing control: Trust and its substitutes' effect on managers involving employees in decision making. *Group & Organization Management, 24,* 155–187.

Thomas, K. W., & Tymon, W. (1994). Does empowerment always work: Understanding the role of intrinsic motivation and interpretation. *Journal of Management Systems, 6,* 84–99.

Thomas, K. W., & Velthouse, B. A. (1990). Cognitive elements of empowerment: An "interpretive" model of intrinsic motivation. *Academy of Management Review, 15,* 666–681.

Weick, K. E., & Sutcliffe, K. M. (2007). *Managing the unexpected: Resilient performance in an age of uncertainty* (2nd ed.). San Francisco, CA.: Jossey-Bass.

Wraith, R. (1994). The impact of major events on children. In R. Watts & D. J. de la Horne (Eds.), *Coping with trauma.* Brisbane, Australian Academic Press.

Chapter 10

PROTECTING THE PROTECTORS: THE RESILIENCY INTEGRATION MODEL

JOHN M. VIOLANTI

INTRODUCTION

What have not been fully explored in police work are factors that provide a protective or buffering effect between work stressors and personal stress. One factor is resiliency, the adaptive capacity (Paton, Violanti, & Lunt, 2010) of individual officers and/or entire organizations to maintain balance in the presence of stress. Our previous work has suggested that the police organization is not only considered a difficult work stressor by officers, but that it also does not provide the resilient structure necessary for officers to deal with daily stressors or traumatic incidents (Paton et al., 2008). Antonovsky (1990) suggested that resilience reflects the extent to which individuals can call on psychological and physical resources that allow them to render stressful events coherent and manageable. In this chapter, we propose that a police officer's ability to render stress manageable reflects the interaction of *person and organization,* which provides a proactive framework for sustaining resilience. This may lead to a reduction of the pathogenic consequences of stress as well as related sequalae such as sick leave, disability, and injury.

Police work produces conditions of both intense and chronic stress. During violent incidents, split-second judgment is critical to avoid unnecessary harm to the public, to coworkers, and to self. Officers know that these incidents may lead to litigation, second-guessing of officer decisions, and loss of personal property in court settlements, further increasing the stressful milieu. Police officers (861,000 in the United States) are also exposed to physical harm, shift work, long work hours, organizational stressors, victims of violence, police suicide, and other tragic events. These occupational environmental challenges contribute to an increased physical and psychological stress load.

If one views the impact of police stress from a pathological perspective, the research presently suggests negative health outcomes. For example, Violanti, Vena, and Petralia (1998) suggested that police officers die an average age at death 10 years younger than that of the general U.S. population. Police deaths are also associated with higher prevalence of cardiovascular disease (CVD) morbidity. This health disparity increases as officers age and retire. After controlling for the influence of traditional CVD risk factors, it nearly doubles by the sixth decade and exceeds a twofold increased risk by the seventh decade (Franke, Ramey, & Shelley, 2002; Ramey, Downing, & Franke, 2009). Higher prevalence of CVD difference is found in younger officers using early subclinical markers of CVD (Joseph et al., 2009, 2010). This disparity in health for police officers is thought to be related to the higher levels of stress exposure inherent in police work (Joseph, et al, 2010; Ramey, et al, 2009).

Judging from this and other research, police work is generally viewed as a precursor to the development of pathological psychological and physiological stress outcomes. However, little work has explored possible salutogenic protective influences that can help officers become more resistant to the impact of stress and reduce the incidence of stress-related health outcomes. Antonovsky (1993) posited that a salutogenic orientation versus a pathogenic orientation has far wider implications than simply proving a directive to focus on the health rather than the pathology of stress. According to Antonovsky (1987; 1990), a "sense of coherence," or a way of making sense of the world, is a major factor in determining how well a person manages stress and stays healthy. This is in some ways similar to the treatise of Janoff-Bulman (1992) in *Shattered Assumptions* and of meaning discussed by Herman (1992). Antonovsky (1993) suggested that the neurophysiological, endocrinological, and immunological pathways through which the sense of coherence operates influence health outcomes associated with stress.

Recent evidence suggests that resilience has a moderating influence on stress (Aldwin, Levenson, & Spiro, 1994; Paton, Violanti, & Smith, 2003). Resilience is often used to imply an ability to "bounce back" from adverse events. There are specific characteristics that make individuals or organizations more or less resilient to stress than others (Luthar, Chicchetti, & Becher, 2000; Maddi, 2002; Maddi & Khoshaba, 2005; Walsh, 2002; Youssef & Luthans, 2005). Taken together, these factors suggest that resilience is not so much a trait as it is a process (Siebert, 2002), although physiology and genetics certainly may play a role in resilience (Baker, Risbrough, & Schork, 2008; Friedl & Penetar, 2008; Yehuda, 2004). Evidence indicates that people can also learn to be resilient through past experience and by developing qualities that facilitate coping, adaptation, and recovery from stress (Luthans, Vogelgesang, & Lester, 2006).

There are different approaches to resilience. Maddi and Khoshaba (2005) examined resilience as hardiness and concluded that three factors form the basis of psychological resilience: commitment, control, and challenge. Walsh (2002) studied resilience from a family and group dynamics perspective and identified four processes essential to group resilience: (1) the group's belief system, (2) its organizational pattern, (3) its specific communication process, and (4) its broader interaction pattern (see Lukey & Tepe, 2008). Coutu (2002) argues that the capacity to improvise can be found in organizations (Lukey & Tepe, 2008). A positive organizational climate empowers employees to exercise judgment to do whatever it takes to get the job done right and on time. Adaptive processes in collaborative problem solving play a role in allowing resilient organizations to buffer stress and to manage a threatening environment effectively. Resilient organizations possess improvisational flexibility or might develop the capacity over time in response to stressful circumstances (Lukey & Tepe, 2008; Turnispeed, 1999).

As Siebert (2002) has suggested, resiliency is best described as a process. Police officers respond to stressful incidents as members of agencies whose climate influences their thoughts and actions (Paton, Smith, Ramsay, & Akande, 1999; Weick & Sutcliffe, 2007). The organization can influence the individual, and organizations can change the course of individual reaction from pathogenic decline to adaptation. Higgins (1994) suggested that coping style and social cohesion could act to cognitively integrate the stressful experience. The salutogenic effects of resilience and social perception suggest that the group can facilitate the active process of self-righting and growth. Police organizational structure also has an influence on how the organization responds to challenges as well. The size, type, and configuration of police districts may be more or less appropriate for the demands of the environment at a particular time. Other structural considerations include the workload of a specific district, how it is staffed, and the ratio of leaders to officers.

In sum, it is important to examine police stress and trauma from a salutogenic rather than a pathogenic perspective. A salutogenic orientation has far wider implications than simply proving a directive to focus on the health rather than the pathology of stress. The positive integration of organization and individual in police work can have a meaningful impact on increasing officer resilience and the more effective management of stress. In the long term, reduced personal stress has the potential for positive health outcomes in this difficult occupation. Figure 10.1 represents a proposed model, the Resiliency Integration Model, which can be operationalized using existing measures found in the literature (Paton et al., 2008).

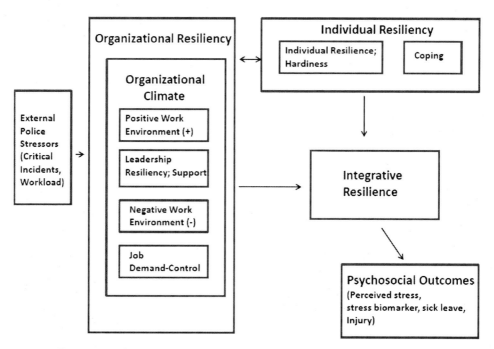

Figure 10.1. Stress and Organizations-Individual Resilience Integration.

OPERATIONAL SUGGESTIONS FOR
THE RESILIENCY INTEGRATION MODEL

Traumatic Incidents at Work

Traumatic events can be measured using the Police Incident Survey developed by Violanti and Gherke (2004). The questions were extracted from a survey of all traumatic events police listed and are aimed at eliciting information about the type and frequency of disturbing incidents such as officer-involved shootings the officers encountered at work during the past 12 months.

Organizational Climate

There are several factors that make up organizational climate: (1) work environment (positive and negative), (2) leadership, and (3) demand- control balance (Paton et al., 2008).

Leadership. The Multifactor Leadership Questionnaire (MLQ-5X) (Avolio & Bass, 2002) may be used to obtain self-ratings of leadership behavior. The MLQ questionnaire contains 45 items describing behavior and attributions, each rated on a five-point scale (0 = rarely, 4 = to a large extent). Three outcome variables can be extracted from this measure; indexing transformational, transactional, and passive-avoidant leadership behavior. Resiliency has been associated with increases in transformational and transactional leadership styles and decreases in passive-avoidant style (Avolio & Bass, 2002). The MLQ-5X will be completed by both leaders (assessing their own attributes) and line personnel (assessing their perceptions of leaders).

Police Daily Hassles and Uplifts Scale. The Police Daily Hassles and Uplifts Scales (Hart, Wearing, & Headey, 1993, 1994). This scale may be used to measure the positive (beneficial to well-being) and negative (harmful to well-being) work experiences encountered by police on a *day-to-day* basis (Paton et al., 2008). The coefficient alphas range from .77 to .93 for the hassles dimensions and from .52 to .92 for the uplifts dimensions.

Demand-Control Balance. How much balance officers have between job demands and control over such demands is an important organizational variable. In the context of the demand-control model (Karasek, Brisson, Kawakami, Houtman, Bongers, & Amick, 1998; Karasek & Theorell, 1990). Resilience might be understood as potentially helpful to the perception of control or use of social support. Specific characteristics of resilience (e.g., willingness to improvise and determination) bear obvious relevance to control and thus may facilitate experience and perception of control (Lukey & Tepe, 2008). The second contribution of this measure is the proposition that social support serves to buffer the expected effects of extreme demands. If this idea is correct, it suggests an additional avenue by which resilient individuals might benefit. Resilient people may be more willing and active in seeking social support (Lukey & Tepe, 2008; Walsh, 2002). The proposition that social support can play a significant role in buffering against strain suggests the need to recognize an additional and potentially important marker of resilience itself. As individuals seek social support, they may strengthen and reinforce their own and others' capacity for resilience (Campbell, Campbell, & Ness, 2008; Lukey & Tepe, 2008).

The Job Content Questionnaire (JCQ) is an instrument based on the demand-control model and is used to measure the psychosocial content of a participant's work situation. The JCQ is based on situations in which high psychological demands in concert with low job control lead to job "strain" (stress), which in turn predicts adverse psychological and physical heath reactions (Karasek & Theorell, 1990). This instrument measures the following constructs: skill discretion, decision authority, macrolevel decision authority, psychological job demands, physical job demands, job insecurity, supervisor social support, and coworker social support (Karasek, 1985).

Individual Resiliency

Hardiness as personal resiliency may be measured using the 15-item Dispositional Resiliency Scale developed by Bartone (1995) consisting of three hardiness dimensions including control, commitment, and challenge. The control dimension consists of items that represent the characteristic of believing that one is capable of managing potentially stressful events (e.g., planning ahead can help avoid most future problems). The commitment dimension consists of items that represent an ability to find meaning in potentially stressful events (e.g., most of my life gets spent doing things that are worthwhile). The challenge dimension has items related to the ability to interpret potentially stressful events as opportunities (e.g., changes in routine are interesting to me). The Connor-Davidson Resilience scale (CD-RISC 10) (Connor & Davidson, 2003) is a second standardized measure of resilience that can be applied to personal qualities. The scale demonstrates that resilience is modifiable and can improve with treatment, with greater improvement corresponding to higher levels of global improvement.

Outcome Measures

Stress. Stress can be measured on several dimensions. Many instruments which measure perceived stress from an individual perspective. A frequently used stress scale is the Perceived Stress Scale, which is a measure of global rather than event-specific stress levels. It is a 10-item self-report symptom inventory designed to measure the degree to which someone identifies situations as being stressful. The Perceived Stress Scale has been found to be internally consistent and is recommended for use when assessing nonspecific stress in relation to disease outcomes or behavioral disorders (Cohen, Kamarck, & Mermelstein, 1983; Hewitt, Flett, & Mosher, 1992).

It is also important to measure stress from a biological perspective. The allostasis model provides a way of dealing with this difficulty by providing a conceptual model that is complex enough to encompass both psychological and biological components of exposure and response to stressors. The human system for responding to environmental challenges involves the central nervous system, the hypothalamic pituitary adrenal (HPA) axis, and the autonomic nervous system (ANS). The inability of this system to appropriately inhibit the sympathetic excitatory stress response and the HPA axis response has been implicated in allostatic load (McEwen, 1998). The terms "allostasis" and "allostatic load" refer to aspects of an integrative biopsychosocial model for stress (McEwen, 2007; McEwen & Wingfield, 2003). According to this model, repeated stress is viewed in terms of the following linked components: (1) external conditions affect an individual, (2) an appraisal of these conditions is made, (3) coping strategies are exercised, and (4) psychological and biological effects are realized after modulation by such factors as resiliency, appraisal and coping. Resiliency and the HPA axis have not been evaluated to any significant degree.

Posttraumatic Stress Disorder (PTSD) Symptoms. There are several excellent measures available for PTSD symptoms. Two commonly used instruments are the Impact of Event Scale-Revised (IES-R) (Weiss, 2004) and the PCL-C (Weathers, Litz, Herman, Huska, & Keane, 1993: Norris & Hamblin, 2004). The IES-R has been noted for its ability to measure change in PTSD symptoms over time and is useful for longitudinal analyses because it is available for all participants in the baseline sample. The Posttraumatic Stress Checklist-Civilian Version (PCL-C) has been noted for its diagnostic and psychometric properties and is useful in cross-sectional analyses where diagnostic categories are of interest (Ruggiero, Del, Scotti, & Rabalais, 2003). The IES-R consists of 22 items describing the subjective impact or symptom related to a traumatic event. These items are related to three response sets or subscales, including Intrusion, Avoidance, and Hyperarousal. Subscales for Intrusion (seven items), Avoidance (eight items), and Hyperarousal (seven items) are obtained by calculating the mean of the appropriate items. The overall IES-R score is obtained by summing all 22 items. The PTSD checklist civilian version (PCL-C) consists of 17 self-report items rated on a five-point scale. Each of the 17 items refers to how much the individual has been bothered by the PTSD symptom listed in the item (e.g., repeated, disturbing memories, thoughts or images of a stressful experience from the past). An overall PTSD symptom severity score is obtained for the PCL-C by summing the items. The PCL-C has also been shown to have high internal consistency, test-retest correlation, and convergent validity (Ruggerio et al., 2003).

FUTURE WORK

The mental, behavioral, and social costs of police stress and trauma suggest a substantial need for intervention (Armeli, Gunthert, & Cohen, 2001; Violanti et al., 2006). Various levels of stress management may be applied in police work. Establishing a "stress shield" (Paton et al., 2008) that provides an integration of organizational and individual resiliency is part of this shield. While tertiary and secondary prevention methods are generally considered, such treatment services are underutilized among police (Foa & Rothbaum, 1998) likely due to a police culture that stigmatizes admission of emotional problems. A more effective approach might be to establish a primary stress prevention and resiliency development protocol for officers. Primary prevention efforts that bolster resilience may be viewed as acceptable to the police culture because preparation for future events fits that standard police model, does not stigmatize officers, and targets maintenance of good mental and behavioral health, rather than treating "mental illness" after it develops (Bartone, Barry, & Armstrong, 2009).

This chapter has described a model of resilience proposing the interaction between person and organizational factors. However, the benefit of any model is a function of it being theoretically rigorous and capable of moving research to practice. The components that we employ in this model are amenable to change through organizational intervention and change strategies. This confers both theoretical rigor and practical utility for first responder organizations and personnel.

Other Suggestions

- Research suggests that organizational influence on resilience is homogenous. However, most organizations consist of lower level subdivisions that have their own characteristics and influences. To assess the differential influence of resiliency throughout a police organization, it may be important to extend analyses to all levels of the organization in order to distinguish between *macrolevel* organizational influences (leadership, police agency rules, regulations, and directives) and *microlevel* influences (leadership, workload, geographic traits, communications, and district policies). At the same time, it is important to recognize that some but not all microlevel policies and procedures are influenced directly by macrolevel policies and standards. It is proposed that levels higher in total resilient characteristics (such as police districts or precincts) will be associated with lower psychosocial pathogenic outcomes in those levels.
- Because police officers are called on repeatedly to deal with increasingly complex and threatening incidents, it is also appropriate to expand the scope of resiliency to include the development of one's capacity to deal with future events (Klien,

Nicholis, & Thomalla, 2003). Each negative stressful event an officer encounters leads to an attempt to cope, which forces him or her to learn about his or her own capabilities and organizational support networks. Thus, it is important to examine integrated organizational and individual experiences over time to assess how the temporal effects of resiliency may be challenged by unpredictable incidents and organizational reactions that occur in policing.

CONCLUSIONS

Although police work is traditionally viewed as a precursor to the development of acute and chronic posttraumatic stress reactions, there is growing evidence that it is associated with adaptive, positive (e.g., resiliency) outcomes (Aldwin, Leveson, & Spiro, 1994; Maddi, 2002). The pathogenic model focuses on the more immediate problem of the individual and appropriate therapy, whereas the salutogenic orientation is concerned with overall mental health and utilizes strengths within the group (Antonovsky, 1993). When one realizes that the outcome of a stressor is not preordained stress, one can see that even undesirable stressors can have salutary outcomes.

Present research does not adequately examine the integration of organizational and individual resiliency in police work. We propose that this interactional influence is more effective in stress management than they are separately. For example, the concept of hardiness in both leaders and subordinates may have a positive influence on the group's ability to manage stress. Hardiness represents the ability of an individual to become accustomed to fatigue or hardship, to be capable of withstanding austere or horrific conditions (Funk, 1992).

An increased understanding of factors in police work that may provide a protective buffer against chronic and traumatic stress may lead to an increased police health impact and future interventions designed to reduce exposure to stressors and enhance coping strategies and resiliency. In addition, such understanding may lead to innovative police organizational strategies to help reduce psychological, physical, and health consequences among officers.

REFERENCES

Aldwin, C. M., Levenson, M. R., & Spiro, A., III. (1994). Vulnerability and re-silience to combat exposure: Can stress have lifelong effects? *Psychology and Aging, 9,* 34–44.

Antonovsky, A. (1987). *Unraveling the mystery of health. How people manage stress and stay well.* San Francisco: Jossey-Bass Publishers.

Antonovsky, A. (1990). Personality and health: Testing the sense of coherence model. In H. Friedman (Ed.), *Personality and disease.* New York: John Wiley and Sons.

Antonovsky, A. (1993). The implications of salutogenesis: An outsider's view. In A. Turnbull, J. Patterson, S. Behr, & D. Murphy (Eds.), *Cognitive coping, families, and disability.* Baltimore, MD: Paul H. Brookes.

Armeli, S., Gunthert, K. C., & Cohen, L. H. (2001). Stressor appraisals, coping, and post-event outcomes: The dimensionality and antecedents of stress-related growth. *Journal of Social and Clinical Psychology, 20,* 366–395.

Avolio, B. J., & Bass, B. M. (2002). *Manual for the Multifactor Leadership Questionnaire (Form 5X).* Redwood City, CA: Mindgarden.

Baker, D. G., Risbrough, V. B., & Schork, N. J. (2008). Posttraumatic stress disorder: Genetic and environmental risk factors. In B. J. Lukey & V. Tepe (Eds.), *Biobehavioral resilience to stress* (pp. 177–218). Boca Raton, FL: CRC Press.

Bartone, P., Barry, C. L., & Armstrong, R. E. (2009, November). To build resilience: Leader influence on mental hardiness. *Defense Horizons.*

Bartone, P. T. (1995). *A Short Hardiness Scale.* Paper presented at the American Psychological Society annual convention, New York, May.

Campbell, D., Campbell, K., & Ness, J. W. (2008). Resiliency through leadership. In B. J. Lukey, & V. Tepe (Eds.), *Biobehavioral resilience to stress* (pp. 57–88). Boca Raton, FL: CRC Press.

Cohen S., Kamarck T., & Mermelstein R. (1983). A global measure of perceived stress. *Journal of Health and Social Behavior, 24,* 385–96.

Connor, K. M., & Davidson, J. R. T. (2003). Development of a new resilience scale: The Connor-Davidson resilience scale (CD-RISC). *Depression and Anxiety, 18,* 76–82.

Coutu, D. (2002, May). How resilience works. *Harvard Business Review,* pp. 46–55.

Foa, E. B., & Rothbaum, B. O. (1998). *Treating the trauma of rape: Cognitive behavioral therapy for PTSD.* New York: Guilford.

Franke W., Ramey S., & Shelley M. (2002). Relationship between cardiovascular dis-ease morbidity, risk factors, and stress in a law enforcement cohort. *Journal of Occupational and Environmental Medicine. 44,* 1182–1189.

Friedl, K. E., & Penetar, D. M. (2008). Resiliency and survival in extreme environ-ments. In B. J. Lukey & V. Tepe (Eds.), *Biobehavioral resilience to stress* (pp. 139–176). Boca Raton, FL: CRC Press.

Funk, S. C. (1992). Hardiness: A review of theory and research. *Health Psychology, 11,* 335–345.

Hart, P. M., Wearing, A. J., & Headey, B. (1993). Assessing police work experiences: Development of the Police Daily Hassles and Uplifts Scale. *Journal of Criminal Justice, 21,* 553–572.

Hart, P. M., Wearing, A. J., & Headey, B. (1994). Perceived quality of life, personality and work experiences: Construct validation of the Police Daily Hassles and Uplifts Scales. *Criminal Justice and Behavior, 21,* 283–311.

Hewitt P., Flett G., & Mosher S. (1992). The Perceived Stress Scale: Factor structure and relation to depression symptoms in a psychiatric sample. *Journal of Psychopathology and Behavioral Assessment 14,* 247–57.

Herman, J. (1992). *Trauma and recovery: Aftermath of violence from domestic abuse to political terror.* New York: Basic Books.

Higgins, G. O. (1994). *Resilient adults: Overcoming a cruel past.* San Francisco: Jossey-Bass.

Janoff-Bulman, R.: Shattered Assumptions: Toward a New Psychology of Trauma. New York: The Free Press, 1992.

Joseph, N.P., Violanti, J.M., Donahue, R., Andrew, M.E., Trevisan, M., Burchfiel, C.M., & Dorn, J. (2009). Police work and subclinical atherosclerosis. Journal of Occupational and Environmental Health, 51, 700-707.

Joseph, P.N., Violanti, J.M., Donahue, R., Andrew, M.E., Trevisan, M., Burchfiel, C.M., & Dorn, J. (2010). Endothelial function, a biomarker of subclinical cardiovascular disease in urban police officers. Journal of Occupational and Environmental Medicine, 52, 1004-1008.

Karasek R. A. (1985). *Job Content Questionnaire and Users Guide.* Lowell, MA: University of Massachusetts.

Karasek R., Brisson C., Kawakami N., Houtman I., Bongers P., & Amick B. (1998).The Job Content Questionnaire (JCQ): An instrument for internationally comparative assessments of psychosocial job characteristics. *Journal of Occupational Health Psychology, 3,* 322–355.

Karasek R., & Theorell, T. (1990). *Healthy work: Stress, productivity, and the reconstruction of working life.* New York: Harper Collins.

Klein, R., Nicholls, R., & Thomalla, F. (2003). Resilience to natural hazards: How useful is this concept? *Environmental Hazards, 5,* 35–45.

Luthans, F., Vogelgesang, G. R., & Lester, P. (2006). Developing the psychological capital of resiliency. *Human Resource Development Review, 5,* 25–44.

Luthar, S., Cicchetti, D., & Becher, B. (2000). The construct of resilience: A critical evaluation and guidelines for future work. *Child Development, 71,* 543–562.

Lukey B. J., & Tepe V. (2008). *Biobehavioral resilience to stress.* Boca Raton, FL: CRC Press.

Maddi, S. (2002). The story of hardiness: Twenty years of theorizing, research, and practice. *Consulting Psychology Journal, 54,* 173–185.

Maddi, S., & Khoshaba, D. (2005). *Resilience at work: How to succeed no matter what life throws at you.* New York: Amacom.

McEwen, B. S. (1998). Protective and damaging effects of stress mediators. *New England Journal of Medicine, 338,* 171–179.

McEwen, B. S. (2007). Physiology and neurobiology of stress and adaptation: Central role of the brain. *Physiology Review, 87,* 873–904.

McEwen B. S., & Wingfield J. C. (2003). The concept of allostasis in biology and biomedicine. *Hormone Behavior, 43,* 2–15.

Norris F. H., & Hamblin J. L. (2004). Standardized self-report measures of civilian trauma and PTSD. In J. P. Keane & T. M. Wilson (Eds.), *Assessing psychological trauma and PTSD* (2nd ed., pp. 63–102). New York: Guilford.

Paton, D., Smith, L. M., Ramsay, R., & Akande, D. (1999). A structural re-assessment of the Impact of Event Scale: The influence of occupational and cultural contexts. In R. Gist & B. Lubin (Eds.), *Response to disaster.* Philadelphia, PA: Taylor & Francis.

Paton, D., Violanti, J. M., Johnston, P., Burke, K. T., Clarke, J., & Keenan, D. (2008). Stress shield: A model of police resiliency. *International Journal of Mental Health, 10,* 95–108.

Paton, D., Violanti, J. M., & Lunt, J. (2010). Developing resilience in high risk professions: Integrating person, team, and organizational factors. In B. Pattanayak, P. Niranjana, K. Ray, & S. Mishra (Eds.), *Storming the global business: The rise of the tigers* (pp. 303–316) New Delhi, India: Excel Books.

Paton, D., Violanti, J. M., & Smith, L. M. (2003). Resilience and growth in high-risk professions: Reflections and future directions. In D. Paton, J. M. Violanti, & L. M. Smith (Eds.), *Promoting capabilities to manage posttraumatic stress: Perspectives on resilience* (pp. 204–209). Springfield, IL: Charles C Thomas.

Ramey, S. L., Downing, N. R., & Franke, W. D. (2009). Milwaukee police department retirees: Cardiovascular disease risk and morbidity among aging law enforcement officers. *AAOHN, 57,* 448–453.

Ruggiero K. J., Del B. K., Scotti J. R., & Rabalais A. E. (2003). Psychometric properties of the PTSD Checklist–Civilian Version. *Journal of Traumatic Stress, 16,* 495–502.

Siebert, A. (2002, July). How resilience works. Letter to editor. *Harvard Business Review,* p. 121.

Turnipseed, D. (1999). An analysis of the influence of work environment variables and moderators of the burnout syndrome. *Journal of Applied Social Psychology, 25,* 782–800.

Violanti, J. M., & Gehrke, A. (2004). Police trauma encounters: Precursors of compassion fatigue. *International Journal of Emergency Mental Health, 6,* 75–80.

Violanti, J. M., Vena, J. E., Burchfiel, C. M., Sharp, D. S., Miller, D. B., Andrew, M. E., Dorn, J., Wende-Wactaski, J., Beighley, C. M., Pierino, K., Joseph, P. N., & Trevesan, M. (2006). The Buffalo Cardio-Metabolic Occupational Police Stress (BCOPS) Pilot Study: Design, methods, and measurement. *Annals of Epidemiology, 16,* 148–156.

Violanti, J. M., Vena, J. E., & Petralia, S. (1998). Mortality of a police cohort: 1950–1990. *American Journal of Industrial Medicine, 33,* 366–373.

Walsh, F. (2002). A family resilience framework: innovative practice applications. *Family Relations, 51,* 130–139.

Weathers F. W., Litz B. T., Herman D. S., Huska J. A., & Keane T. M. (1993). *The PTSD checklist (PCL): Reliability, validity, and diagnostic utility.* Paper presented at the annual meeting of the International Society for Traumatic Stress Studies, San Antonio, TX.

Weick, K. E., & Sutcliffe, K. M. (2007). *Managing the unexpected: Resilient performance in an age of uncertainty* (2nd ed.). San Francisco, CA: Jossey-Bass.

Weiss D. S. (2004). The impact of events scale–revised. In J. P. Wilson & T. M. Keane (Eds.), *Assessing psychological trauma and PTSD* (2nd ed., pp. 168–189). New York: Guilford.

Yehuda, R. (2004). Risk and resilience in posttraumatic stress disorder. *Journal of Clinical Psychiatry, 65* (Suppl. 1).

Youssef, C. M., & Luthans, F. (2005). Resiliency development of organizations, leaders & employees: multi-level theory building for sustained performance. In W. Gardner, B. J. Avolio, & F. O. Walumbwa (Eds.), *Authentic leadership theory and practice. Origins, effects, and development.* Oxford, UK: Elsevier.

INDEX

Charles C Thomas
PUBLISHER • LTD.

P.O. Box 19265
Springfield, IL 62794-9265

- Freeman, Edith M.—**NARRATIVE APPR-OACHES IN SOCIAL WORK PRACTICE: A Life Span, Culturally Centered, Strengths Perspective.** '11, 218 pp. (7 x 10), 7 il., 19 tables.

- Lamis, Dorian A. & David Lester—**UNDER-STANDING AND PREVENTING COLLE-GE STUDENT SUICIDE.** '11, 336 pp. (7 x 10), 21 il., 11 tables.

- Voris, Steven J.—**DEVOTIONS AND PRAY-ERS FOR POLICE OFFICERS: Providing Meaningful Guidance for Police Officers. (2nd Ed.)** '11, 194 pp. (7 x 10).

- Wadeson, Harriet C.—**JOURNALING CAN-CER IN WORDS AND IMAGES: Caught in the Clutch of the Crab.** '11, 210 pp. (7 x 10), 70 il. (includes a CD-ROM—"Cancer Land: An Altered Book for an Altered Life").

- Douglass, Donna—**SELF-ESTEEM, RECO-VERY AND THE PERFORMING ARTS: A Textbook and Guide for Mental Health Practitioners, Educators and Students.** '11, 258 pp. (7 x 10), 6 il., 5 tables, $55.95, hard, $35.95, paper.

- Violanti, John M., Andrew F. O'Hara & Teresa T. Tate—**ON THE EDGE: Recent Perspectives on Police Sui-cide.** '11, 158 pp. (7 x 10), 2 il., 4 tables, $39.95, hard, $24.95, paper.

- Anthony, Kate, DeeAnna Merz Nagel & Stephen Goss — **THE USE OF TE-CHNOLOGY IN MENTAL HEALTH: Applications, Ethics and Prac-tice.** '10, 354 pp. (7 x 10), 6 il., 5 tables, $74.95, hard, $49.95, paper.

- Hendricks, James E., Jerome B. McKean, & Cindy Gillespie Hendricks —**CRISIS INTER-VENTION: Contemporary Issues for On-Site Interveners. (4th Ed.)** '10, 412 pp. (7 x 10), 4 il., 1 table, $77.95, hard, $55.95, paper.

- Bryan, Willie V.—**THE PROFESSIONAL HEL-PER: The Fundamentals of Being a Helping Professional.** '09, 220 pp. (7 x 10), $51.95, hard, $31.95, paper.

- Emener, William G., Michael A. Richard & John J. Bosworth—**A GUIDEBOOK TO HUMAN SER-VICE PROFESSIONS: Helping College Students Explore Opportunities in the Human Services Field. (2nd Ed.)** '09, 286 pp. (7 x 10), 2 il., 4 tables, $44.95, paper.

- Stepney, Stella A.—**ART THERAPY WITH STUDENTS AT RISK: Fostering Resilience and Growth Through Self-Expression. (2nd Ed.)** '09, 222 pp. (7 x 10), 16 il., (14 in color), 19 tables, $56.95, hard, $38.95, paper.

- Violanti, John M. & James J. Drylie—**"COPI-CIDE": Concepts, Cases, and Controversies of Suicide by Cop.** '08, 130 pp. (7 x 10), 2 tables, $36.95, hard, $21.95, paper.

- France, Kenneth—**CRISIS INTERVENTION: A Handbook of Immediate Person-to-Person Help. (5th Ed.)** '07, 320 pp. (7 x 10), 3 il., $65.95, hard, $45.95, paper.

- Violanti, John M. & Stephanie Samuels —**UNDER THE BLUE SHADOW: Clinical and Behavioral Perspectives on Police Suicide.** '07, 192 pp. (7 x 10), 1 il., $34.95, paper.

5 easy ways to order!

 PHONE: 1-800-258-8980 or (217) 789-8980

 FAX: (217) 789-9130

 EMAIL: books@ccthomas.com
Web: www.ccthomas.com

 **MAIL:** Charles C Thomas • Publisher, Ltd. P.O. Box 19265 Springfield, IL 62794-9265

Complete catalog available at ccthomas.com • books@ccthomas.com

Books sent on approval • Shipping charges: $7.75 min. U.S. / Outside U.S., actual shipping fees will be charged • Prices subject to change without notice